Systems Mapping

"Just about everywhere you look there are examples of interconnected networks of nodes, whether they be causal factors, people, organisms, or actors. The dynamic relationships between these define everything from how ecosystems function to political, social and financial networks. They are the stuff of both Nature and Society, and they often display complex, non-linear behaviour. If we are ever to be able to constructively manage these systems, we first need to be able to define their state and this involves a process of system mapping. This book is a welcome practical guide to the ways in which it is possible to map systems. Anybody who manages systems—and that involves almost all of us—will benefit from the insights provided in this book."

—Professor Sir Ian Boyd FRS, *Professor of Biology* (*University of St Andrews*) and former Chief Scientific Advisor at *Defra*

"Those of us engaged with the complexity frame of reference in science have long recognized the need for the development of methods which put it to work. This is not just or even primarily for scientific investigation but even more importantly for addressing the interwoven social and environmental challenges facing us—crises is not too strong a word—and developing policy and practice to get things that have to be done, done. Systems mapping is a developing set of techniques, participatory/co-production in character, which provide ways of doing just that. The authors of this book, drawing on a long experience of practice informed by complexity thinking, give us what is not only a well thought out account of the approaches but also a practical manual for using them. Systems mapping is a combination of science, art and practical skill. It has enormous potential and this book will play an important part in getting people to use it and use it well."

—Professor David Byrne FAcSS, *Emeritus Professor of Sociology and Applied Social Science* (*Durham University*)

"I thought I knew a thing or two about system diagrams … I now know an awful lot more. More importantly, I know I'm in safe hands when authors write about their own experiences so openly and freely. So it is here in this excellent, accessible, practical, readable and comprehensive book. I felt like I was being taken on a journey with a really committed and experienced set of tour guides."

—Bob Williams; Author of System Diagrams; a Practical Guide. https://gum.co/systemdiagrams

"'Systems Mapping: how to build and use causal models of systems' says it all. If you want to understand, in very pragmatic and practical ways, what this approach is all about and how to use it, then this is the book for you. Written by two of the leading experts in the field, Systems Mapping is based on the valuable lessons they have learned over the years—including which techniques work (or do not work) for a given situation and how and why, which is massively helpful. While systems mapping tools are easy to run, co-producing a system map in practical, actionable, and participatory ways can be challenging at times, and rightly so—their purpose is to help us, as stakeholders, come to some agreement on how best to understand and improve the complex systems problems we presently face, from the environment and economy to government and public policy. I highly recommend this book and will use it in my classes and policy evaluation workshops, as it offers a powerful approach for making sense of today's complexity, but in a way that differs from and yet adds to the existing repertoire of computational, statistical, historical, and qualitative methods."

—Professor Brian Castellani, PhD, FAcSS, Director of the *Durham Research Methods Centre* and Co-Director of the *Wolfson Research Institute for Health and Wellbeing, Durham University*

"Pete Barbrook-Johnson and Alex Penn have written the right book at the right time; finding their moment as the demand grows for making sense of pervasive complexity. The book is intensely pragmatic, informed by practitioners who have been 'swimming in shark-infested waters' as they describe it. The reader is eased in with welcome clarity and honesty, teeing up seven accessible method-specific chapters. Three further cross-cutting issues chapters treat the reader to wholly pragmatic insights on data and evidence; running a mapping process; and comparing, choosing and combining methods to suit the situation. Throughout the book, the authors make clear that systems mapping must, above all, be useful. This book clearly achieves that standard and will create durable value."

—Gary Kass, Deputy Chief Scientist, Natural England; Visiting Professor, *Centre for Environment and Sustainability, University of Surrey*; Vice-President, *Institution of Environmental Sciences*

"Over the last few decades, there has been a major shift in policy thinking towards accepting that the social and economic world is not like a machine following a predictable path, but is complex, with feedback, tipping points and adaptation. With this has come an increasing need for better ways to understand social, economic and political systems as a whole. One of these ways is system mapping, but until now there has not been a comprehensive and easily understood guide about how to do it. This book is pioneering in bringing together a wide range of system mapping techniques, explaining with great clarity how they can be used and where each of them is appropriate. I congratulate the authors on writing a book that will be an invaluable guide for everyone interested in understanding a complex world."

—Professor Nigel Gilbert CBE, ScD, FREng, FAcSS, *Professor of Sociology, University of Surrey. Director of the Centre for the Evaluation of Complexity Across the Nexus (CECAN)*

Pete Barbrook-Johnson • Alexandra S. Penn

Systems Mapping

How to build and use causal models of systems

Pete Barbrook-Johnson
School of Geography and the
Environment
University of Oxford
Oxford, UK

Alexandra S. Penn
Department of Sociology
University of Surrey
Guildford, Surrey, UK

ISBN 978-3-031-01833-6   ISBN 978-3-031-01919-7   (eBook)
https://doi.org/10.1007/978-3-031-01919-7

This Palgrave Macmillan imprint is published by the registered company Springer Nature Switzerland AG.
The registered company address is: Gewerbestrasse 11, 6330 Cham, Switzerland

ACKNOWLEDGEMENTS

Pete would like to thank colleagues at CECAN and the University of Surrey, especially Alex, for the intellectual encouragement, stimulus, and support to pursue research on and using systems mapping. He would also like to thank colleagues at the Institute for New Economic Thinking and the School of Geography and the Environment at the University of Oxford for the encouragement and freedom to continue this line of research and for time to work on this book. Finally, he would like to thank Jen, Billy, Mart, and Doug for their love and support, and dedicates his part of this book to them.

Alex would like to thank all colleagues, especially project partners and workshop participants, who have helped to develop our thinking on PSM and system mapping in general within the discipline of practice. In particular, Martha Bicket, Jean Davies, Adam Hejnowicz, Caitlin Jones, Verena Knerich, Chris Knight, Fran Rowe, Helen Wilkinson, Anna Rios-Wilks, and Beth Wills. Particular thanks also go to Kasper Kok for introducing Alex to system mapping, being such a generous collaborator, and exemplifying the participatory spirit! Thanks go to Colin for humour, support, and wise advice. And most particularly to Pete for his enthusiasm, energy, intellect, and friendship over not just this project but many years of collaboration.

We would both like to thank CECAN Ltd (www.cecan.co.uk), the consultancy-arm of CECAN, and especially Nigel Gilbert, for generous financial support towards producing this book.

Finally, we would like to thank Dione Hills and Helen Wilkinson (ToC), Philippe Vandenbroeck (CLD), Simon Henderson and Stuart

Astill (BBN), Stephen Morse (RP), Kasper Kok (FCM), and Birgit Kopainsky (SD) specifically for the illuminating conversations they held with us about the methods in this book. We hope to release these conversations as podcasts/interviews via CECAN in the coming months. They all gave their time, expertise, and experience so generously. This book would not have been possible without them. Their openness stands as a real testament to the values and worldview which underpin participatory modelling approaches.

About the Book

There is a growing need in a range of social, environmental, and policy challenges for a richer more nuanced, yet actionable and participatory, understanding of the world. Complexity science and systems thinking offer us hope in meeting this need, but in the past have often only offered either (i) highly technical 'black-box' modelling, (ii) appealing metaphors and language which don't directly lead to action, or (iii) overwhelming and paralysing complexity.

Systems mapping is a front runner in meeting this need, providing a key starting point and general-purpose resource for understanding complex adaptive systems in practical, actionable, and participatory ways. However, there is confusion about terms and methods, an underappreciation of the value they can bring, and a fundamental underestimation of the differences between approaches and the resulting outputs of mapping processes and analysis.

This book explores a range of new and older systems mapping methods focused on representing causal relationships in systems. In a practical manner, it describes the methods and considers the differences between them; describes how to use them yourself; describes how to choose between and combine them; considers the role of data, evidence, and stakeholder opinion; and describes how they can be useful in a range of policy and research settings. The book focuses on practical insights for causal systems mapping in real-world contexts, with tips from experienced practitioners, and a detailed guide on the realities and challenges of building and using these types of system maps.

CONTENTS

About the Authors

Pete Barbrook-Johnson is Departmental Research Lecturer in the Economics of Environmental Change in the Environmental Change Institute (ECI) and the Smith School for Enterprise and the Environment, both in the School of Geography and the Environment at the University of Oxford. He is also a member of the Institute for New Economic Thinking at Oxford and a research associate at St Catherine's College.

Pete's core research interests sit at the crossroads of social science and economics, complexity science, and environmental and energy policy. He uses a range of methods in his research including agent-based modelling, network analysis, and systems mapping. He regularly uses these, and other methods, to explore applied social, economic, and policy questions, and to support complexity-appropriate policy evaluation, but is equally interested in more theoretical aspects of complex adaptive systems.

Pete teaches on a range of undergraduate and masters' courses across the School of Geography and the Environment, specialising in the economics of environmental change, and the use of complexity and systems sciences in environmental issues.

He has conducted research with and for the likes of UK government departments/agencies such as Defra, BEIS, the Environment Agency, and the Health and Safety Executive, and businesses such as Anglian Water and Mott Macdonald. Internationally, he has collaborated with research institutes and government in China, South Africa, Italy, and Ethiopia.

Pete is also a member of the Centre for the Evaluation of Complexity Across the Nexus (CECAN) and a Visiting Fellow at the Centre for Research in Social Simulation (CRESS) and Department of Sociology at the University of Surrey. He sits on the editorial boards of both the *International Journal of Social Research Methodology* and *Humanities and Social Science Communications*. Previously, Pete was a UKRI-ESRC Innovation Fellow and senior research fellow working on public-private partnerships and collaboration, a 'Knowledge Integrator' in CECAN, a research fellow at the Policy Studies Institute, and a PhD student and then research fellow at CRESS. Prior to his PhD, Pete studied Economics at the University of East Anglia, before completing his MSc in Environmental Technology (specialising in Environmental Economics and Policy) at Imperial College London.

Pete is on twitter @bapeterj and his personal website is https://www.barbrookjohnson.com/

Alexandra S. Penn is a complexity scientist working on combining participatory methodologies and mathematical models to create tools for stakeholders to understand and 'steer' their complex human ecosystems. As a senior research fellow at the University of Surrey, she has developed participatory complexity science methodologies for decision makers to explore interdependencies between social, ecological, economic, and political factors in 'industrial ecosystems', in particular, looking at the transition to bio-based economy in a region of heavy industry and fossil fuel energy generation in the Humber Estuary, UK. She is a principal member of CECAN, a collaboration between academics, policy professionals, and the UK government to generate novel, cutting-edge methods for evaluating policy for complex systems.

Alex has an academic background in physics and evolutionary ecology, training at Sussex University and as a junior fellow at the Collegium Budapest Institute for Advanced Study, followed by a Life Sciences Interface fellowship in the Science and Engineering of Natural Systems Group, University of Southampton. She is also a strong inter-disciplinarian, with a track record of working across disciplines, with a broad variety of stakeholders from policymakers to industrialists and with members of the public as a science communicator.

She was made a fellow of the Royal Society of Arts for her work in the novel application of whole-systems design to bacterial communities, is a frequent visiting researcher at ELSI, the Institute for Earth and Life Sciences at Tokyo Tech University, and is Societal Impact Editor of the *Journal of Artificial Life* and Associate Editor of the *Journal of Adaptive Behaviour*.

LIST OF FIGURES

LIST OF TABLES

CHAPTER 1

Introduction

Abstract This chapter introduces the book and the topic of systems mapping. We explain our motivation for writing the book, what 'systems mapping' means to us, our focus on causal approaches, and what methods are included, and which are not, in the book. We also explore how these methods are related to one another. We begin to consider how systems mapping can be useful in research and practice, before making the case for why we believe it is worth thinking about now.

Keywords Systems mapping • Modelling • Complexity • Policy
• Systems

This book introduces systems mapping and outlines seven methods that allow us to develop causal models of systems. We focus on the practical realities of how and when to use these methods and consider wider issues such as what types of evidence and data to use in their construction, how to run workshops, and how to compare, choose, and combine methods. We do not cover all types of systems mapping, we almost entirely ignore those which do not focus on cause and influence in systems, nor do we delve into the deeper philosophical ideas underpinning their use.

Writing this book feels a bit like swimming in shark-infested waters. Not least because several people have told us that is indeed what we are

© The Author(s) 2022
P. Barbrook-Johnson, A. S. Penn, *Systems Mapping*,
https://doi.org/10.1007/978-3-031-01919-7_1

doing! Some of these systems mapping methods, and the underlying ideas, have been around for some time and there are many people with strong views on them. Despite this, we believe there is much confusion around these methods. There is an underappreciation of their value, but also the large differences between methods. We do not wish to attempt to declare for once and for all what should or should not be called 'systems mapping', nor offer the definitive definition of any specific method. But we do hope to make the landscape of methods clearer, to help people find, understand, and use these methods more easily.

Our paths to systems mapping were not straightforward, neither were they similar. Pete was looking to broaden his methodological expertise after spending nearly six years using agent-based modelling in academic research; he wanted to find methods which were more accessible and usable in a range of contexts, that were less reliant on lots of data for validation, or lots of time or money to do. Whereas, Alex, moving from the natural to the social sciences, but with experience in participatory systems design, was looking for participatory methods that could be used in projects taking a complex systems approach. She was also looking for approaches that could work quickly under the pressure of expectant project partners, without empirical data, and with a large multi-disciplinary team.

Our relationships with systems mapping since have also not been simple. We have become frustrated at times, but we have always found ourselves drawn back, either through our own intellectual curiosity (or inertia!), or through the needs of stakeholders and research users. What has been consistent throughout is the ability of systems mapping approaches to provide us with academically stimulating ideas and to do this in an intuitive way which generates usable and timely insights, and value to the people we work with.

Why Did We Write This Book and Who Is It For?

At times, it has been hard to work with systems mapping. Some people see it as one simple method and miss the wealth of different approaches and what they can do. Others see the detail of one or two approaches and go deep into only those. Systems mapping is also often subsumed into the world of 'systems thinking', somewhat hidden by that wider philosophy on how to understand, be, and act in the world.

Put simply, systems mapping is a hard space to navigate. As we learnt and applied our knowledge, we often felt a little lost, without the right tools to guide us. This book is an attempt to solve that. In one sense, it is written for our former selves; it is what might have helped us accelerate our learning and practice more quickly.

It is also hopefully for you. For people who are thinking that systems mapping might be useful in their work but who are not sure where to start. Or people who want to use a particular method but need to ground this in a wider context, need some help to get started, and don't want to miss any opportunities to do it well. It is also for those who are familiar with one method but who would like an overview of what others exist or might be useful in different problem contexts. Or for those who have heard about systems mapping and would like to get a sense of what it is about.

This book is intentionally practical and pragmatic. We are not preaching from the 'High-Church of systems mapping' but pounding the streets. We are looking for ways forward, trying to shine some light on dark alleys, looking for ways to improve ourselves. This introductory chapter asks, 'what is systems mapping?' and 'why look at it now?' and tries to be honest about the breadth and noise in the answers to these questions. From here, we embark on seven mini-adventures, exploring systems mapping methods in detail.

WHAT IS SYSTEMS MAPPING?

Let's be honest, systems mapping means lots of different things; it is broad and ill-defined. We are not going to 'fix' that here (if we even think it needs fixing). We support inclusive and broad definitions in general, and think they are inevitable when it comes to systems mapping. But that breadth and inclusivity should not come at the cost of clarity. We still need to know where we are at, and what is on either side of us.

In time-honoured academic fashion, let's start by breaking this down into its component parts, and first asking what is a 'system'? There is no simple answer to this question. We regularly see arguments about whether something is a system or not, whether a system mapping exercise has taken enough care thinking about what the system it is mapping, or even whether we should be mapping problems not systems at all. While these concerns are important, it is possible to define almost anything as a system with enough mental gymnastics. Moreover, what the 'right' system definition for you is will always be context dependent. This means we would rather proceed with thinking about what your system is, rather than dwelling on what a system is. Your system might be the system you are part of, and you wish to understand, or the system which you are going to map, hopefully with some purpose in mind.

Nonetheless, given that others have considered what a system is, it's worth looking at a couple of our favourite definitions. Williams and Hummelbrunner (2011) suggest that there are a few distinctions we can all agree on: (i) that systems are made up of some set of elements; (ii) that systems also constitute the links between elements, whether they are processes or interrelationships; and (iii) that systems have some boundary, and this is central to their definition. They accept, as we do, that this set of distinctions could mean almost anything, so they suggest focusing on what is distinctive about seeing the world with a systems lens, rather than dwelling on definitions.

Meadows (2008) takes this definition one step further, bringing in the ideas of purpose and organisation, suggesting a system is an 'interconnected set of elements that is coherently organised in a way that achieves something' (pg. 11). The idea of a system purpose, and using it to help define your system, and perhaps your mapping exercise, is useful but slippery. It will likely require you to have a broad definition of a purpose, to include functions, services, or value that a system may provide.

The second component of 'systems mapping' is 'mapping'. So, what is a 'map'? Here we bump into an unfortunate historical quirk of terminology. In the systems mapping world, 'map' is used synonymously with 'model'. They are both reasonably intuitive words, but there has been a lot of thought about what a model is, and separately, what a map is, some of which has ideas in common, but plenty which does not. Maps are normally thought of in the cartographic, geographic sense, a representation of a physical space. There is fascinating literature on considering what these types of maps are and how they shape our thinking. Some of this is useful when thinking about models and system maps, but some of it is a distraction.

More useful, we think, is the history of thought on modelling and, within this, asking 'what is a model?' As with systems, there are many definitions and types of model, but there is a little more consistency and a settled general definition. We would characterise this definition as this: a model is a purposeful simplification of some aspect or perception of reality. 'All models are wrong, but some are useful' (Box and Draper, 1987) is the modelling cliché to end all modelling clichés, but it is instructive. The simplifications a model makes in its representation of reality mean it is inherently 'wrong' (i.e. it is not reality), but if these simplifications serve some purpose, then there is a decent chance that they are useful.

So now we know what a system is, and what a map is. Do we know what a system map is? Not quite. We should stop talking in the abstract and show you with examples, but first we need to introduce a few common components of systems maps:

- **Network**: in its simplest sense, a network is a set of boxes connected by lines. In system maps, these lines are often directed, that is, they are arrows from one box to another.
- **Nodes**: the 'boxes' in a network are normally referred to as nodes.
- **Edges**: the connections, lines, or arrows between boxes are normally referred to as edges.

All except one of the methods in this book always have a network of nodes and edges, representing cause and influence between factors in a system, at their core. These networks of cause and influence are the model (i.e. the map) of the system.

WHAT SYSTEMS MAPPING METHODS ARE IN THIS BOOK?

There are seven systems mapping methods that we go into detail on; they all focus on, or at least allow us to consider, causal patterns. In alphabetical order, here are brief introductory descriptions of each:

1. **Bayesian Belief Networks**: a network of variables representing their conditional dependencies (i.e. the likelihood of the variable taking different states depending on the states of the variables that influence them). The networks follow a strict acyclic structure (i.e. no feedbacks), and nodes tend to be restricted to maximum two incoming arrows. These maps are analysed using the conditional probabilities to compute the potential impact of changes to certain variables, or the influence of certain variables given an observed outcome. These maps can look relatively simple, but they have numbers in, and if you don't like probability, you might not like them.
2. **Causal Loop Diagrams**: networks of variables and causal influences, which normally focus on feedback loops of different lengths and are built around a 'core system engine'. Maps vary in their complexity and size and are not typically exposed to any formal analysis

but are often the first stage in a System Dynamics model. These are popular, and you have likely seen one before.

3. **Fuzzy Cognitive Mapping**: networks of factors and their causal connections. They are especially suited to participatory contexts, and often multiple versions are created to capture diverse mental models of a system. Described as 'semi-quantitative', factors and connections are usually given values, and the impacts of changes in a factor value on the rest of the map are computed in different ways.

4. **Participatory Systems Mapping**: a network of factors and their causal connections, annotated with salient information from stakeholders (e.g. what is important, what might change). Maps tend to be large and complex. Analysed using network analysis and information from stakeholders to extract noteworthy submaps and narratives.

5. **Rich Pictures**: a free-form drawing approach in which participants are asked to draw the situation or system under consideration as they wish, with no or only a handful or gentle prompts. This method is part of the wider group of Soft Systems Methodologies.

6. **System Dynamics**: a network of stocks (numeric values for key variables) and flows (changes in a stock usually represented by a differential equation), and the factors that influence these. Normally, these maps are fully specified quantitatively and used to simulate future dynamics. This is a popular method with a well-established community.

7. **Theory of Change maps**: networks of concepts usually following a flow from inputs, activities, outputs, and outcomes to final impacts. Maps vary in their complexity and how narrowly they focus on one intervention and its logic, but they are always built around some intervention or action. Maps are often annotated and focused on unearthing assumptions in the impact of interventions.

This is not an exhaustive list of system mapping methods—far from it. This list reflects our preferences and biases, and our intention of exploring methods which represent causality and influence in a system, and methods which can be used in a participatory way. Below, we list some of the methods which we do not include in this book, but which are nonetheless potentially useful and relevant for you.

How Do These Methods Relate to One Another?

Let us now consider some of the broad characteristics of the methods that we focus on and how they fit together. To do this, we use three related conceptual spaces in Figs. 1.1, 1.2, and 1.3 and position the methods within these; one on their overall focus and nature (Fig. 1.1), second on their mode and ease of use (Fig. 1.2), and third on the outputs and analysis they produce (Fig. 1.3). It is important to note that these placements are debatable and could misrepresent individual projects' use of a method. However, we believe they give a rough sense of where these methods sit in relation to one another, and more importantly, what some of the most important axes on which to differentiate them are.

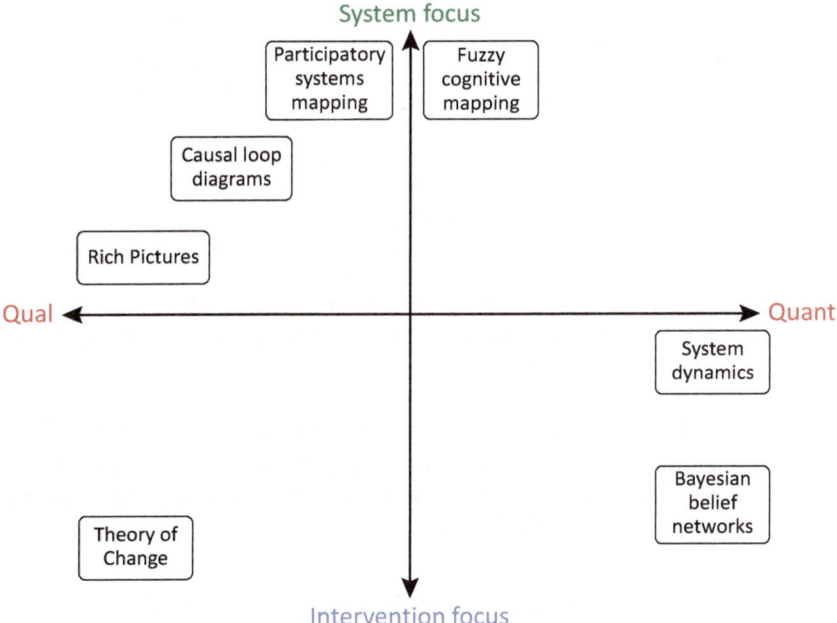

Fig. 1.1 The methods in this book placed on a 'system focus—intervention focus' axis (i.e. does the method emphasise more focus on the whole system or on an intervention), and a 'qualitative—quantitative' axis. Source: authors' creation

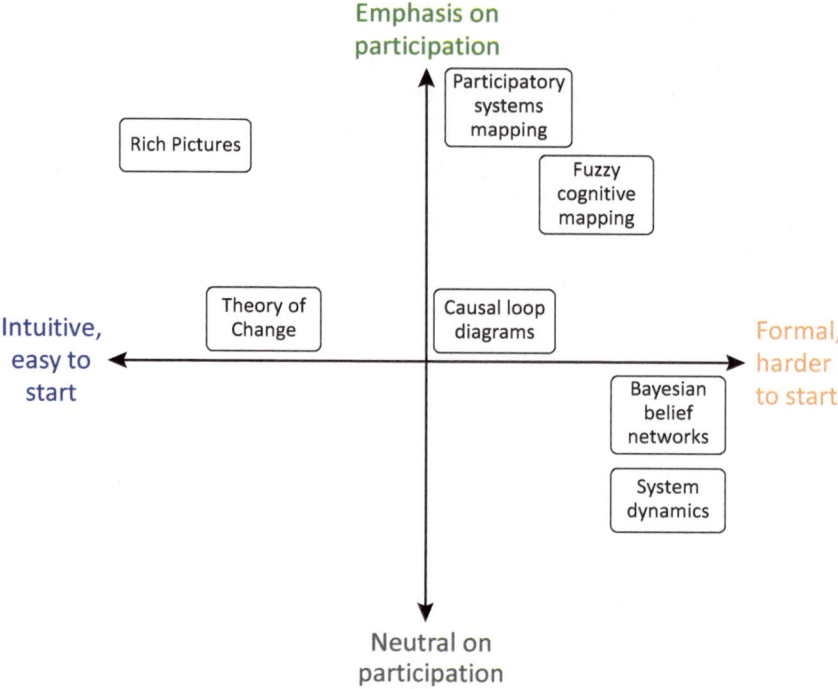

Fig. 1.2 The methods in this book placed on an 'emphasis on participation' spectrum, and an 'intuitive, easy to start—formal, harder to start' spectrum. Source: authors' creation

Figures 1.1, 1.2, and 1.3 give us a quick sense of where the methods sit, but it is possible to elaborate on this further; in Table 11.1 on comparing, choosing, and combining methods, we do this by describing these distinctions, and a few more, in further detail.

WHAT METHODS ARE NOT IN THIS BOOK?

Because of our focus on methods that consider causality in a system, there are many methods which can be classed as 'systems mapping' which we do not include. This does not mean they are not important, or that we do not value them. Below we attempt to outline those we are aware of and point you in the direction of useful resources.

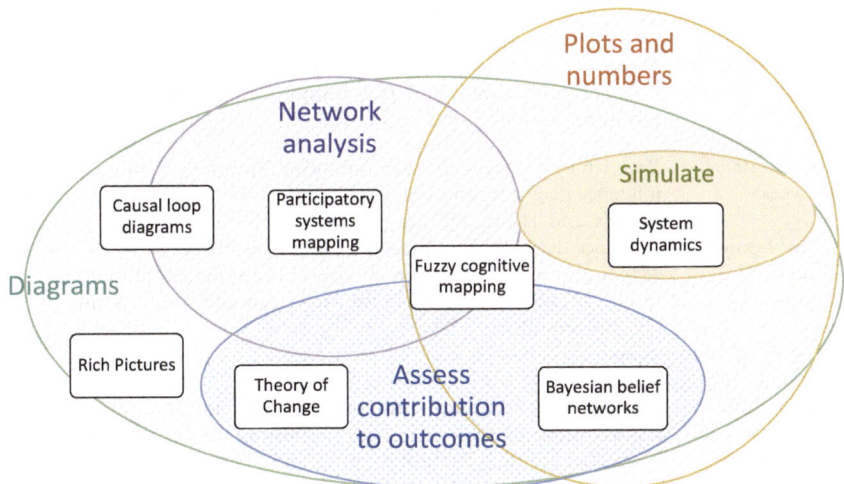

Fig. 1.3 The methods in this book positioned in a Venn diagram by the types of outputs and analysis they produce. Source: authors' creation

Before that, it is worth pointing out some of the different terminology and names that are used elsewhere for methods that are in this book; a selection is listed in Table 1.1. You might find yourself looking for these in this book and being disappointed not to find them—fear not, they are here, just under a different name. There are also a few terms, similar to 'systems mapping', that get used in a loose way and can refer to almost any of the methods in this book, such as 'mind mapping', 'cognitive mapping', 'causal mapping', or 'causal diagram'. We dare not try to unpick the various uses and history of these terms in detail; suffice to say, when you read them elsewhere, make sure to check what they are referring to.

Now let's turn to the methods not included in this book. Table 1.2 overviews these with a brief description, explanation of why they did not meet our criteria, and where you can find more information.

Beyond individual methods or suites of methods, there are several overarching schools of practice, or research sub-disciplines, which offer potential value for systems mapping. We do not cover these in this book because they are covered well elsewhere and they are entire ways of understanding and acting in their own right, not specific mapping methods. Nonetheless, they contain many techniques, tools, and approaches with much in

Table 1.1   Different terminology for the methods covered in this book

Name used in the book	Other terms sometimes used to refer to the method (note, there is an overlap between these and, in fact, many of the terms are also separate methods in their own right)
Bayesian Belief Networks	Bayesian networks, probability networks, dependency models, influence diagrams, directed graphical models, causal probabilistic models, and Theory of Change maps.
Causal Loop Diagrams	Influence diagrams, system maps, sign graphs, Participatory Systems Mapping. You may also see these referred to as System Dynamics models because of their use in the early stages of building System Dynamics models.
Fuzzy Cognitive Mapping	None.
Participatory Systems Mapping	None, though there are approaches based on Causal Loop Diagrams that are sometimes referred to as Participatory Systems Mapping.
Rich Pictures	None, though Rich Pictures is part of Soft Systems Methodology, so you may see this used.
System Dynamics	None.
Theory of Change maps	Programme theory, intervention theory, logic mapping, logic models, results chain, and outcome mapping.

common with systems mapping, so you may find them useful to explore for inspiration, both on individual methods and on wider philosophy. They include the following:

- **Participatory (action) research**: there are large literature on participatory research and 'participatory action research' that foreground the participation and co-production of research with communities and stakeholder groups. These contain many dozens of workshop and focus group methods and techniques, many of which are quick and easy to use, which may be of value to you. They also provide the wider framing, philosophy, and motivation on why it is worthwhile approaching topics from a participatory angle. See Cornwall and Jewkes (1995) or McIntyre (2007) for more.
- **Design thinking and methods**: similarly, there is a large literature and practice around design principles, thinking and methods, and applying these to policy issues and other problems beyond the common understandings of product or industrial design. These approaches include numerous methods for structuring thinking and bring peoples' views to bear on an issue. See the UK Policy Lab

Table 1.2 Methods often referred to as systems mapping which are not in this book

Method(s)	Description	Why not in this book?	Further reading or information
Causal (cognitive) mapping, causal mapping, cognitive mapping.	A collection of tightly related methods for building aggregated causal maps, typically from individual primary interview and survey data, or secondary documentary data.	These methods are arguably all indirectly included via Fuzzy Cognitive Mapping. They sometimes emphasise developing representations of individual mental models rather than representations of systems. These are the methods we most likely would have included had we expanded our focus.	Laukkanen and Wang (2015), Ackermann and Alexander (2016), Axelrod (1976).
(Group) Concept mapping	A method for organising and visualising concepts and ideas among a group of people.	Not focused on a causal understanding of a system.	Kane and Trochim (2007)
Cultural-Historical Activity Theory (a.k.a Activity Theory, Activity Systems, CHAT)	A detailed systems approach, coming from a cognitive psychology starting point, which focuses on learning and the interaction between peoples' feelings and beliefs and their environment.	Not focused on causal understanding of systems. A broader approach.	Williams (2021) for introduction. Foot (2014)
Cynefin	A decision support approach that facilitates exploration and appraisal of different responses or action in situations. Known for the famous 'complex, complicated, chaotic, clear, confusion' quadrant diagram.	Not focused on causal understanding of systems. Broader approach to action in systems.	Williams (2021) for introduction.

(continued)

Table 1.2 (continued)

Method(s)	Description	Why not in this book?	Further reading or information
Giga-mapping	An inclusive approach to mapping the relationships and entities and processes in a system, often with very complex diagrams. Roots in systems-oriented design (SOD)	Not focused on causal description alone.	https://www.systemsorienteddesign.net/index.php/giga-mapping
Log frames, logical frameworks	Used to describe a general approach and specific matrix technique for planning and evaluating projects.	Typically not depicted with networks, but matrices/tables. Similar to Theory of Change in some ways.	https://www.betterevaluation.org/en/evaluation-options/logframe
Mind mapping	Can refer to a range of different types of processes and diagrams, but typically involves relatively free-form connection of entities, processes, and concepts in a radial or tree-like structure.	Not focused specifically on cause.	Buzan (2006)
Outcome mapping	Used to refer to a range of processes and diagrams which connect interventions with their outcomes in a similar way to Theory of Change and log frames.	Often interchangeable with Theory of Change.	https://www.betterevaluation.org/en/plan/approach/outcome_mapping
ParEvo	Participatory method for developing stories of past histories or future scenarios using tree-like diagrams of sequences of events.	Focused on stories and narratives rather than causal models.	https://parevo.org/

Path analysis	An early (1920s) approach to describing the dependencies between variables graphically.	Focused on visual representation of statistical analysis rather than causal modelling.	Wright (1934)
Participatory mapping	A range of methods which develop geographical maps of places in participatory ways to represent the spatial knowledge of people.	Not causal.	Corbett (2009)
Social network analysis	Method for representing and analysing social connections using networks.	Not causal.	Knoke and Yang (2008)
Spray diagram	Generic approach to showing connections between elements or concepts related to an issue. Often in a radial or tree-like structure.	Not focused on causal relationships.	https://www.open.edu/openlearn/science-maths-technology/engineering-technology/spray-diagrams
Stakeholder/actor mapping	Range of approaches to visualising or listing stakeholders/actors in a systems and attributes and/or connections between them.	Not causal.	Too many equally valid references to pick one—a simple search will return many results.
Viable systems model	A systems approach which explores minimum requirements for a system (often some form of collective action, e.g. an organisation) to maintain or produce itself, using diagrams.	Not focused on causal modelling, broader approach to topic of viable systems.	Williams (2021) for an introduction.

(https://openpolicy.blog.gov.uk/) and their Open Policy Making Toolkit for more.

- **Permaculture/systems design:** related to design thinking above are specific schools of thought which focus on the participatory design and management of whole complex systems, their components, and interactions between them for sustainability. Although these approaches are most often applied to the design of geographically located systems, in particular socio-ecological systems with an emphasis on agroecology, they include many useful generalisable tools. The design philosophy of working with systems and the design cycle and process and systems mapping methods used within are particularly useful and have inspired our thinking. See Holmgren (2002) and https://knowledgebase.permaculture.org.uk/design

How Can Systems Mapping Be Useful?

The range of systems mapping methods, from those which are infinitely flexible to those which emphasise participation, those which discipline thinking, and those which allow calculation and simulation, hints at the plethora of ways in which systems mapping can be useful. There is no generic quick answer as to why you would use systems mapping, how it would generate value, and be useful to you. Rather, there is a long list of answers which depend on the context of the system or issue you are working on—your goals, needs, skills and capacity—and whether you are generating value from the process of mapping, from just the end product, or both. This list tends to revolve around five broad types of use, which also apply to most types of modelling or analysis. They are:

1. **Helping us think:** system maps of all types force us to be more specific about our assumptions, beliefs, and understanding of a system. At the very least they force us to 'put it down on paper'. Many types of systems mapping also force us to structure our ideas using some set of rules or symbols (i.e. creating boxes and lines to represent concepts and their relationships). This will introduce simplifications and abstractions, but it will also make explicit our mental models. This, often simple, process disciplines our thinking and exposes it to scrutiny, even if it is only the scrutiny of our own reflections and the structure imposed by the method. Helping us to think

is the most fundamental value systems mapping brings. You basically cannot avoid having it do this to you!

2. **Helping us orient ourselves:** a systems mapping process will often also help us orient ourselves to a system or issue. This is where the word 'map' is particularly apt. Whether a map helps us see our, and others', positions in the system, or whether it helps us quickly develop a fuller understanding of an issue, we will be better oriented to it. This helps people navigate the system better, be aware of what else to think about when considering one part of a map, or know who is affected and so should be included in discussions.

3. **Helping us synthesise and connect information:** the more flexible types of mapping are particularly good at bringing together different types of data, evidence, and information. They can all be used to inform the development of a map, making connections that would not otherwise be possible. Different types of visualisation, hyperlinking, and map structure can also be used to help people return to the information underlying a map.

4. **Helping us communicate:** whether we build maps in groups, or alone, and then share them, all system maps should help us communicate our mental models and representations of systems. This is an often-underestimated benefit of mapping in groups; the process of mapping with others, and the discussions it generates, unearths a multitude of assumptions which can then also be challenged and unpicked. The richness and depth of discussion, while maintaining structure and focus, is often a surprise to first-time participants. The end product of a mapping process can also help us communicate our ideas about a system. Maps can become repositories for our knowledge which can be accessed again and again by others, and updated, becoming a living document. However, it is worth noting that system maps are sometimes referred to as 'horrendograms', and much worse (!), when they show us the complexity of a system in an unfiltered manner. People think in different ways, and there are many people who prefer to use more structure or simplification to communicate or learn. There are cases in which system maps can be unhelpful communication tools if used naively. We say 'naively' because there are many ways, within each method, to avoid this, and to help people 'enter' a map, build understanding, and navigate a potentially overwhelming systems map.

5. **Helping us extrapolate from assumptions to implications:** systems mapping approaches which can be turned into simulations, or which can be analysed in a formal way, also allow us to follow through from the assumptions we have embedded in them, to their implications. The most obvious example is System Dynamics, which allows us to simulate the dynamics of a system. In effect, this allows us to attempt to look forward, to see how the structure and assumptions we have created play out over time. Using models in this way, to 'predict' or 'forecast', is generally well understood, but people sometimes think of systems mapping as more static and are unable to do this. In a related but different way, Bayesian Belief Networks allow us to follow through the implications of the many conditional dependencies we embed in them, to consider what impact a change might bring, or what contributed to an observed outcome. Other approaches provide ways to consolidate and sense check the combined and often contradictory effects of multiple influences on distal factors. Whether by computing numerical values representing potential combined effects of change on outcomes in relative terms (e.g. Fuzzy Cognitive Mapping), or by visualising causal pathways between a changed factor and outcomes, allowing us to think through the multiple indirect effects (e.g. Participatory Systems Mapping).

WHY THINK ABOUT SYSTEMS MAPPING NOW?

The systems and complexity sciences have been around since at least the mid twentieth century, arguably longer, and many of the methods in this book have also been around a decent while. Interest in these ideas and approaches, and attempts to apply them to real-world concerns, has come in waves over the last seventy or so years. There has been notable success but also false dawns, and plenty of scholars and practitioners have been sceptical about their value. In the past, the complexity and systems sciences have sometimes offered either highly technical 'black-box' modelling, appealing metaphors, and language which don't directly lead to action and are often misapplied, or overwhelming and paralysing complexity. These are serious problems, which many are now seeking to address, including us.

Despite these issues (and though we may be biased and myopic), we have observed a renewed interest in the last ten years or so and noted many others making similar observations. It does not feel too outlandish to claim that we are at a high point of interest currently. We bump into fewer and fewer people who have not heard about these ideas, and more and more people actually approach us about them. This is the time of greatest opportunity but also the point at which failure to deliver, or failure to move beyond previous high-water marks, or past pitfalls, may see interest decline rapidly. There is still plenty of confusion and varied use of terminology, arguments over concepts, and underwhelming applications of methods, which can trip us up.

In the context of this current interest in systems thinking and complexity, systems mapping approaches, particularly causal mapping, are particularly useful 'gateway' tools. They can relatively quickly and straightforwardly capture some of the features of complex systems that matter on the ground when trying to understand and manage these systems. In particular, multi-causality, indirect effects, the uncertain boundaries of open systems, feedbacks, and multiple stakeholder perspectives. However, other important complex system characteristics, such as emergent effects, need other modelling approaches. Systems mapping methods are highly usable, useful, and relatively intuitive ways to start engaging with real-world complex systems.

This book represents an attempt to help open up and organise (causal) systems mapping, such that people finding themselves carried along on this wave of interest have something solid to grasp onto and build from. To abuse the metaphor a bit more, we hope when the wave inevitably recedes, more of these ideas and methods, and most importantly the people who believe in them, have got a foothold on the beach and so are not dragged back. We also hope the book helps readers ensure the quality of their use and critique of these methods, so that we see fewer misguided, naïve, or poorly framed applications, and more innovation and combination in their use.

Finally, we hope the book will help users of these methods to navigate one of the biggest headwinds to their success; the increasingly fast-paced nature of work, research, and policy, and the increasing attention deficit of stakeholders and users. It used to be the case that you could organise a workshop over two days, and muddle your way through more easily, learning and adapting a method as you went. Now, if you are lucky, you get a half day of people's time, and since the pandemic, you may only have people's attendance virtually. This puts more pressure on these methods, and this means we need to be better prepared and more efficient at using them.

What's in the Rest of This Book?

Chapters 2 through to 8 cover the seven methods we dive into real detail on, they are roughly in order of the most qualitative through to the most quantitative. We try to build detailed but clear descriptions of what they are and how you can use them, but also reflect on what they are good and bad at, and how things can go wrong. Each of these chapters can be taken on its own, ignoring the rest of the book.

The three chapters after these are more cross-cutting. Chapter 9 considers how and what different types of knowledge and evidence can be used in systems mapping. Chapter 10 dives into the nuts-and-bolts practicalities of running workshops. Chapter 11 considers how we can compare, choose, and combine the methods in this book. Finally, Chap. 12 concludes, with a few final take-home messages, and our reflections on what we have learnt writing this book.

We hope you enjoy it and find it useful. We're always happy to talk systems mapping and get feedback, so feel free to get in touch.

References

Ackermann, F., & Alexander, J. (2016). Researching complex projects: Using causal mapping to take a systems perspective. *International Journal of Project Management, 34*(6), 891–901. https://doi.org/10.1016/j.ijproman.2016.04.001

Axelrod, R. (1976). *Structure of decision: The cognitive maps of political elites.* Princeton Legacy Library. https://press.princeton.edu/books/hardcover/9780691644165/structure-of-decision

Buzan, T. (2006). *Mind mapping.* Pearson Education.

Corbett, J. (2009). *Good practices in participatory mapping: A review prepared for the International Fund for Agricultural Development.* International Fund for Agricultural Development. https://agris.fao.org/agris-search/search.do?recordID=GB2013201933

Cornwall, A., & Jewkes, R. (1995). What is participatory research? *Social Science & Medicine, 41*(12), 1667–1676. https://doi.org/10.1016/0277-9536(95)00127-S

Foot, K. (2014). Cultural-historical activity theory: Exploring a theory to inform practice and research. *Journal of Human Behavior in the Social Environment, 24*(3), 329–347. https://doi.org/10.1080/10911359.2013.831011

Holmgren, D. (2002). *Permaculture: Principles and pathways beyond sustainability.* Holmgren Design Services.

Kane, M., & Trochim, W. M. (2007). *Concept mapping for planning and evaluation.* Sage Publications.

Knoke, D., & Yang, S. (2008). *Social network analysis*. Sage.

Laukkanen, M., & Wang, M. (2015). *Comparative causal mapping: The CMAP3 method*.   Routledge.   https://www.routledge.com/Comparative-Causal-Mapping-The-CMAP3-Method/Laukkanen-Wang/p/book/9780 367879655

Meadows, D. (2008). *Thinking in systems: A primer*. Chelsea Green Publishing.

McIntyre, A. (2007). *Participatory action research*. Sage. https://us.sagepub.com/en-us/nam/participatory-action-research/book230910

Williams, B. (2021). *Systems diagrams: A practical guide.*. https://bobwilliams.gumroad.com/l/systemdiagrams

Williams, B., & Hummelbrunner, R. (2011). *Systems concepts in action: A practitioner's toolkit*. Stanford Business Books.

Wright, S. (1934). The Method of Path Coefficients. *Annals of Mathematical Statistics, 5*, 161–215.

CHAPTER 2

Rich Pictures

Abstract This chapter introduces readers to Rich Pictures and briefly to Soft Systems Methodology, the broader approach from which Rich Pictures emerged. It overviews what Rich Pictures are, how to use them and the steps in doing so, and the common issues and tricks and tips to overcome them. We also consider what Rich Pictures are good and bad at, give a brief overview of their history, and highlight resources and ways to get started.

Keywords Rich Pictures • Soft Systems Methodology

Humans have likely been drawing pictures for as long as we have existed. Free-form visual representation of things, ideas, and processes are universal in human culture and feel like one of the most natural and intuitive ways of expressing ourselves. We draw before we write. It is thus not surprising that drawing pictures can be a useful way of describing, sharing understanding of, and analysing systems. In this chapter we describe and explore the use of 'Rich Pictures' as a systems mapping method.

There is a slight tension in our focus on Rich Pictures. The method comes from the wider approach known as 'Soft System Methodology'. While we will discuss Soft Systems Methodology briefly, our focus is on Rich Pictures alone. Some researchers and practitioners that use Soft Systems Methodology may feel it is inappropriate to take this approach.

© The Author(s) 2022
P. Barbrook-Johnson, A. S. Penn, *Systems Mapping*,
https://doi.org/10.1007/978-3-031-01919-7_2

However, we believe that Rich Pictures, although an outlier in this space, are worthy of discussion as a systems mapping method in their own right. While Soft Systems Methodology as a whole is something much larger which does not easily fit into our definition of what can be considered systems mapping.

The primary reason we chose to include Rich Pictures was that they complete our spectrum of systems mapping methods, from the most formal and quantitative, through more flexible, semi-quantitative, and qualitative approaches, to Rich Pictures, an almost completely free-form approach, with the most flexibility, and which puts all the power and decisions in stakeholders' hands. We felt it was important to have this option in our systems mapping armoury/sewing kit. Rest assured, just because this method is one of the most flexible and free form, it does not mean there are no guidelines for its use and fierce methodological debate around how it should be used.

As in other chapters, we use a simple and practical structure to describe Rich Pictures, starting with as clear and jargon-free description of what the method is, as we can muster. We then describe how to do it, common issues, and tricks of the trade. Next, we step back and consider what the method is good and bad at, before closing with a discussion of the history of the method and pointing out some useful resources for getting started.

WHAT ARE RICH PICTURES?

Rich Pictures are a drawing, a picture, of a system or 'situation'. They are almost always produced together in groups in workshop settings, with large pieces of paper (though some scholars have suggested they can be used as individual analytic tools, e.g. Bell and Morse, 2013a). They are intended to be a shared representation of the system; the value they generate is often mostly in the process and discussions this generates rather than the picture itself as an output. What the picture should contain is often left completely up to the participants; very few, if any, prompts are given by facilitators beyond asking them to 'draw the system'. However, some guidance on the method does suggest that using the prompts, 'structures', 'processes', 'climate', 'people', 'issues expressed by people', and 'conflict', as things to consider putting in the picture can be helpful. Another common prompt for groups who are struggling to start is to suggest they draw themselves in the system. Participants are normally discouraged from using text or words as much as possible, though this is not always the case—some Rich Pictures contain a lot of text. Though not used as a prompt, most facilitators also don't intend participants to produce a diagram which looks like a Theory of Change map or flow diagram. The aim is to avoid the constraints such diagram types introduce. However, Rich Pictures can contain arrows and represent causal relationships. In sum, Rich Pictures are flexible, can contain almost anything, and emphasise letting participants do what they want above all else. Once a Rich Picture is produced, it can be analysed by participants and researchers as part of the process of using the method, though it is often the process of drawing and discussing that is the most important element.

Let's look at some examples. Figure 2.1 shows an 'archetypal-if-poor' Rich Picture of the National Health Service in the UK from Bell and Morse (2013a). We can see the participants who drew the picture in the centre, surrounded by different elements of the system, such as patients and staff (the figures on the right), and concepts such as bureaucracy and

Fig. 2.1 A Rich Picture of the National Health Service in the UK (Source: Bell and Morse, 2013a)

measurement/targets (the abacus and paperwork top left). Bell and Morse describe the picture as being relatively poor in terms of its visual content but suggest that this did not diminish its value as a discussion tool.

Another example can be seen in Fig. 2.2, again from Bell and Morse (2013a). This example benefits from some more skilled drawing perhaps (e.g. the 'see no evil, hear no evil, speak no evil' monkey) and uses no text, except for the 'WB' to denote 'World Bank'.

There is inevitably a huge variety in Rich Pictures, so we strongly suggest you look for more examples to fully appreciate the range in what they

Fig. 2.2 A Rich Picture of the influence of indicators on sustainable development in Slovakia (Source: Bell and Morse, 2013a)

can look like. A simple search engine image search can help with this, or Bell et al. (2016a) includes many examples. Beyond the variety in what participants might produce, there is also variety in the practice of using Rich Pictures. As we acknowledged in our introduction, some practitioners will only use them as part of a wider process, rather than a standalone systems mapping method. This will affect the way in which they are used, the emphasis put on iterating and returning to the pictures, and the amount of time spent focusing on them alone.

There is also variety in the prompts and facilitation given to participants. Most Rich Pictures will be developed with minimal prompts and will not be developed beyond a simple drawing on paper. However, some will be drawn with stronger guidance on what to include, and maybe use rules such as 'no text'. The pictures may also go through some digitisation and refinement, even with the help of an artist or graphic designer, with the aim being to produce something more lasting which can be shared as a communication tool. Lastly, we have observed no variety in the terminology used to describe Rich Pictures, but it is worth noting that many participatory approaches will involve drawing and sketching of different types, and they will have much in common with Rich Pictures, even if they are not formally coming from a systems perspective or intended to 'map' a system in some way.

Rich Pictures emerged from and are part of a wider approach to studying and acting in systems, called Soft Systems Methodology. We do not intend to go into any depth on this approach in this chapter but do outline some of its history and aims in the 'brief history' section below.

How Do You Do Rich Pictures?

The steps in using Rich Pictures are relatively obvious and intuitive. Though they can be tailored to different project's needs, they will typically include the following stages.

- **Planning:** you will need to decide who to invite to a Rich Picture workshop(s) and how to structure the sessions. You may want to do some pre-workshop work on deciding the focus or definition of what system will be looked at. If you are working with a particular client or project partner, they will be key in making decisions at this stage.
- **Workshop:** an individual workshop can be done quite quickly, in as little as thirty minutes if needed, but more commonly around two

hours. The number of participants at a workshop can be relatively large, perhaps as many as twenty per facilitator, and with whole groups as large as fifty. However, each group drawing a picture should be smaller, around three to six people. As the facilitator you will need to decide what prompts, if any, you want to use with participants. This will depend on your own preferences and style but also the purpose of the process and anything specific you intend to do by way of analysing the pictures. Common decisions include (i) whether to ban or discourage the use of text; (ii) whether to mention the list of prompts—structures, processes, climate, people, issues expressed by people, and conflict; and (iii) whether to encourage participants to draw themselves in the picture. In contrast with some of the other methods in this book, we would suggest taking as minimal a role as possible. The method works best when participants are comfortable, and it is likely that too much guidance will disrupt their creativity and expression. Some practitioners even advocate leaving the room during the main drawing time, to avoid the chances of participants asking for help they don't really need, or the temptation for you to hover over those drawing. The final element of the workshop will involve reporting back to the whole group what small groups have drawn. Ideally, this should not just be a short section tagged on the end of a workshop, but should involve at least one cycle of groups sharing what they have done, hearing from others, and then going back to their Rich Picture, updating it, and then sharing again. The discussion within small groups, and between them, as they share their approach and views, is likely to be equally important, if not more important, than the picture itself. You can be creative in designing the process of drawing, discussing, drawing new pictures, or updating existing ones to suit your needs. Ideally, some fieldnotes should be made of the discussions, so that you have a record. It is often impractical to record discussions with an audio recording device, and it may inhibit participants from speaking freely. More likely to be of value would be asking participants to take some notes, or have some observers take notes. A choice will need to be made about whether to take fieldnotes of all discussions or only the full group.

- **Analysis:** although not always done, it is common to do some form of analysis on the pictures generated. This can be started during the workshop discussions, and then continued by the practitioner or researcher afterwards. The purpose and nature of the analysis will depend on the purpose of the project, but it can range from simple narrative and thematic analysis and comparison of the pictures (including reference to discussions during the workshop) through to more formal aesthetic analysis of the images and what this might convey (e.g. as in Bell and Morse 2013a), or structured content analysis of the pictures (see Bell et al. 2016b), using the types of social research methods used for analysing documents.

COMMON ISSUES AND 'TRICKS OF THE TRADE'

During a workshop there are two common issues that can arise which we would like to highlight. Firstly, participants can think that drawing is childish in some way or not valuable and can thus be hesitant to contribute or be sceptical about the method/process. This can be a particular issue for Rich Pictures, compared to other methods in this book, because they do not have the immediate 'feel' of being a practical tool, or a scientific modelling method. Often, any scepticism is overcome with a little time and the influence of positive engagement of others. However, users of the method should think ahead about how they might assuage concerns along these lines. As a facilitator you want to have enough legitimacy and credibility that people want to take part, but not so much that participants think you should have all the answers or are afraid to express ideas in front of you.

Secondly, power dynamics or dominant individuals can influence the picture and its content strongly. Individuals can force a group to draw only their view, or others may be too nervous or fearful to contribute. Because the method is so free-form and flexible, and we normally avoid prompting too much, there is little scope for using the excuse of 'the method says we should do X or should include person(s) Y more' with Rich Pictures. Thus, consideration and management of power dynamics and dominant individuals can only be done in the planning and inviting stages.

Once we have some Rich Pictures and are carrying out some analysis, it is common for those new to the method to struggle to develop rich analyses. People can feel unsure of what analytical tools to use, what can and can't be inferred or said, or how to connect the pictures to other parts of a project. This is normal, and developing rich nuanced and sensitive analysis is difficult and takes time, both within one project and across multiple projects—you will learn and improve a lot in the analysis you do as you do more.

Finally, we do sometimes observe quite serious 'research fatigue' in participants who have taken part in participatory research in the past and not seen tangible results, or those who have simply been involved in many projects. This is a very real issue for any participatory method or project but can be more acute with Rich Pictures because the method is open and flexible. It does not impose a structure on people which may make them feel this is a 'new' or different process, without an immediate instrumental value, and it empowers participants meaning that they can express their fatigue more quickly.

There are a range of useful tips and tricks to deal with these issues and others. Some of the most useful we have come across include:

- **Use icebreakers:** it is a good idea to have a handful of icebreaking suggestions to help participants get through blocks related to scepticism, feeling drawing is childish, feeling they cannot draw well, etc. For any block you think participants might have, arm yourself with an icebreaker. One of the most used for people who are struggling to start (for any reason) is to ask them to draw themselves first (this can induce much laughter, quite literally breaking a static atmosphere) and then build from there. For sceptical participants, open an honest discussion about the use and value of what you are doing, and show them you have their concerns in mind and are not naïve about what is useful or what is a sensible use of their time.
- **Give power to the participants:** do everything and anything you can to hand over your power as the facilitator to the participants. Encourage them and emphasise the value of their opinions and knowledge. Leave the room entirely during the drawing stage (if you need to stay, avoid hovering nearby, explain that you don't want to inhibit them if needed). Make sure they describe their picture first before you or others comment on it. Giving away your power here takes courage as facilitator but is vital to this method.
- **Don't try to force-fix issues during a session:** it can be tempting to try to 'fix' group issues as they emerge by more strongly facilitating group dynamics or what is being drawn. This is almost always a bad idea with Rich Pictures. People will understand that the method is about flexibility and may interpret your attempts as critiques that what they are doing is wrong in some way. Aim to adapt the overall process in planning stages rather than let knee-jerk reactions during a workshop drive your management of the process.

WHAT ARE RICH PICTURES GOOD AND BAD AT?

We hope it is clear from our description thus far, Rich Pictures' strength lies in its flexibility and openness, meaning a process can evolve in almost any direction, and that the method can be bent to almost any purpose. Indeed, the method can easily be bent to the will of participants, it does not constrain them or force them to adopt a modelling framework, meaning it can focus on what is important to them. The method is excellent at quickly opening lively discussions, drawing on humour and expression to help participants develop richer shared understandings of an issue. Visual metaphors (such as the puppet master example in Fig. 2.2) are powerful and quick ways to communicate these understandings. It excels at capturing different perspectives, values, and perceptions often crucial in determining what happens on the ground in social systems, but extremely difficult to capture with formal modelling methods. Rich Picture's flexibility mean they have the potential to allow people to offer whole systems views without constraints or simplifying assumptions. This is a strength, but equally, we should be conscious that they do not enforce or directly encourage a whole systems view, so this does not always emerge.

Rich Pictures are an easy method to start using, there are few resources needed, and though analysis can be difficult to develop quickly, the method itself is not intimidating or technically challenging to use. This means that the barrier to stakeholder participation and engagement is correspondingly low, and most people would be able to contribute their perspectives, including those who might feel intimidated by other methods. It can be used in situations in which participants are not literate, where there are language barriers, or with participants who are unaccustomed to network-type representations. It is also worth noting, Rich Pictures processes rarely fail. Even in a tricky process, or a group the facilitator feels did not work well together, there is still something to work with, some learning to be had, from the discussion and the picture (however simple). Other methods in this book are more likely to fail because of certain essential elements that must be collected or addressed; this is not the case with Rich Pictures.

The flipside of these strengths is the 'weaknesses' of Rich Pictures. We use inverted commas here because these are not really weaknesses, rather just things Rich Pictures will never do because it prizes freedom and expression so highly. The method will not help us formalise knowledge in any precise way, rather it will tend to create discursive and rich descriptions of issues, rather than neater or simplified ones. It will not provide direct inputs into more formal modelling approaches, including those in this book. It may help open a discussion to frame another method, but you cannot use the picture itself to directly seed a Causal Loop Diagram, for example (unless perhaps you strongly facilitate the picture to this end, which would border on just building a Causal Loop Diagram).

A Brief History of Rich Pictures

Rich Pictures emerged as part of the Soft Systems Methodology (SSM). This approach was primarily developed by Peter Checkland at Lancaster University in the UK in the 1970s onwards, as a way of exploring and making decisions in complex 'situations' (Checkland and others seemed to prefer this term to 'systems'—reflecting less emphasis on attempting to describe whole systems) in which there was a lack of agreement on the issue and ways of managing it. The use of 'soft' is in opposition to 'hard' engineering systems in which there is no disagreement on the issue and reflects Checkland's background as a natural scientist and the department at Lancaster he was in—Systems Engineering. The approach developed out of a wider programme of action research and acknowledgement that many projects failed because of issues around agreeing problem definitions. SSM was initially used in organisations as part of management and business research but has now been used in many different settings and domains.

SSM entails a simple and intuitive process of developing an understanding of a system (this is the stage in which Rich Pictures are used), developing options for interventions or management, and then implementing them. Rich Pictures are typically perceived as one of the main innovations of the approach, along with the 'CATWOE' analysis tool. CATWOE stands for 'customers, actors, transformation, worldview, owner, and environment', and comprises a method for analysing issues using these six perspectives, emphasising finding solutions to any issues identified in these domains.

The exact nature of SSM has shifted through the years as it has been used and refined, but detailed descriptions of it, and how to use it, can be found in Checkland and Poulter (2006) and one of the original texts, Checkland (1981). For those interested in the detail of its history, the retrospective discussion in Checkland (2000) is well worth reading. Rich Pictures as a method remained very much part of SSM until more recently. Bell and Morse (2013a) have been arguably most influential in advocating for thinking of, and using, Rich Pictures as a standalone research method. They emphasised the analytic value of the method and its use in wider range of research and practice settings.

Getting Started with Rich Pictures

Diving into using Rich Pictures is relatively easy. You probably could make a good stab at it from reading this chapter alone. Nonetheless, it is worth sharing some of the most useful resources around which can take you deeper into the method and how to use it. These include:

- Bell et al.'s (2016a) book on Rich Pictures (as a standalone method) is a detailed and creative exploration of its use and value. It is almost certainly worth purchasing if you know you will be using Rich Pictures in your work.
- From the same group of authors there are also a string of academic papers which approach the method in a similar way and are well worth reading. Bell and Morse (2013b) provide a detailed example of using Rich Pictures while also introducing and advocating for the method. Bell and Morse (2013a) provide a more general exploration of Rich Pictures and how they can be used in a range of ways, providing useful discussions of their history and how to approach analysis. Finally, Bell et al. (2016b) look in detail at how the tools of content analysis can be applied to Rich Pictures.
- On Soft Systems Methodology there are multiple 'original' and new texts which are useful; we would recommend Checkland (1975) for one of the earliest descriptions of the approach, Checkland (1981), the earliest book on the topic, Checkland (2000) for a detailed retrospective discussion on the approach, and Checkland and Scholes (1990) for an update with more case studies and examples. For those wanting even more, the special issue of the journal *Systems Practice and Action Research* (issue 13 from 2000), in celebration of Checkland's seventieth birthday, may appeal.

As with many of the methods in this book, we would advocate just diving in and using Rich Pictures. You can use them in your team at work, or with family and friends, to test-run it, and start to think about how you might use and tailor the method for your needs. Probably the most important, or rather most difficult thing to dive into quickly, is how to go about analysing Rich Pictures. To consider this more deeply we recommend looking at examples from academic papers (including those above), and thinking about how different documentary, aesthetic, and social research methods could be applied to Rich Pictures.

REFERENCES

Bell, S., & Morse, S. (2013a). How people use rich pictures to help them think and act. *Systemic Practice and Action Research, 26*, 331–348. https://doi.org/10.1007/s11213-012-9236-x

Bell, S., & Morse, S. (2013b). Rich pictures: A means to explore the "sustainable mind"? *Sustainable Development, 21*(1), 30–47. https://doi.org/10.1002/sd.497

Bell, S., Berg, T., & Morse, S. (2016a). *Rich pictures: Encouraging resilient communities.* Routledge. https://doi.org/10.4324/9781315708393

Bell, S., Berg, T., & Morse, S. (2016b). Rich pictures: Sustainable development and stakeholders - The benefits of content analysis. *Sustainable Development, 24*(2), 136–148. https://doi.org/10.1002/sd.1614

Checkland, P. (1975). The development of systems thinking by systems practice—A methodology from an action research program. In R. Trappl & F. Hanika (Eds.), *Progress in cybernetics and systems research* (Vol. II, pp. 278–283). Hemisphere.

Checkland, P. (1981). *Systems thinking, systems practice.* Wiley, Chichester.

Checkland, P. (2000). Soft systems methodology: A thirty year retrospective. *Systems Research and Behavioral Science, 17*(S1), S11–S58. 10.1002/1099-1743(200011)17:1+<::aid-sres374>3.0.co;2-o.

Checkland, P., & Poulter, J. (2006). *Learning for action: A short definitive account of soft systems methodology and its use, for practitioners, teachers and students.* John Wiley and Sons Ltd.

Checkland, P., & Scholes, J. (1990). *Soft systems methodology in action* (2nd ed.). Wiley.

CHAPTER 3

Theory of Change Diagrams

Abstract This chapter introduces Theory of Change diagrams, a popular approach to mapping the causal logic between interventions, their impacts, and the assumptions they rely upon. Despite a wide variety in practice, we attempt to outline what Theory of Change diagrams are, how you can use them, and where they sit in the wider Theory of Change approach. We describe their strengths, weaknesses, and a brief history, and point readers to useful resources, as well as offer some tips for getting started.

Keywords Theory of Change • Evaluation • Policy • Logic models • Logframe

On a superficial level, we could have included Theory of Change (ToC) in this book simply because it is popular and we often see ToC diagrams that look like the diagrams other systems mapping methods produce. However, ToC is more than 'just' an individual system mapping method; it is also an entire approach in its own right to framing, structuring, and implementing the design and evaluation of interventions. ToC has its roots in the evaluation community and literature—that is, the discipline focused on assessing the impact and success of interventions. From these roots, it has spread and is now also used widely in the design of interventions. In some domains, it has also become a core component in the communication of any project, programme, or organisation; in the same way we might ask

© The Author(s) 2022
P. Barbrook-Johnson, A. S. Penn, *Systems Mapping*,
https://doi.org/10.1007/978-3-031-01919-7_3

what our aims and goals are, we now often ask what is our Theory of Change. In these domains, if you can't quickly explain your ToC, you will often be thought of as ill-prepared or naïve.

The success of ToC in becoming a core component of how we communicate our reasoning with others, and how we frame the design and evaluation of interventions, is founded on the value it has given to people in many different settings. It has, on a basic level, been a very useful tool in many different places. This success is also a reflection of its flexibility and ambiguity. It is used in numerous different ways, which makes it a little difficult to talk about from a methodological perspective. There are many different flavours of ToC, arguably as many as there are people using it, and the term can be used somewhat glibly, or may mean many different things. We won't be able to describe the full variety here. Nonetheless, we were keen to include it in this book for several reasons. Firstly, because of the immense and undeniable value it has given to people. Secondly, because in its purest forms it maintains a dogged focus on using the logic of cause and effect to understand the impacts of interventions in systems. And thirdly, because it has a practicality and immediate instrumental value, it is a useful connection from and complement to some of the more abstract or exploratory methods in this book, through to the harsh realities of doing systems mapping in pragmatic settings such as government, large businesses, or third-sector organisations.

In the rest of this chapter, we focus mainly on ToC diagrams specifically, though we do touch on the broader ideas and interpretations of the ToC approach. We explain what it is and how it is done. We consider common issues, how these are overcome, and some tricks of the trade for creating and using ToC diagrams. We also make some assertions about what it is good and bad at. We finish by giving a brief history of ToC and highlighting some useful resources for those of you thinking about using it.

WHAT IS THEORY OF CHANGE MAPPING?

In this chapter, we want to focus on ToC diagrams, but it is worth making clear ToCs often also come in the form of text. Textual ToCs can be as short as one sentence, or longer, perhaps a paragraph or two, but don't tend to be much longer than half a page or so. We have seen examples of textual ToCs structured in tables, breaking down the elements in a more organised way.

ToC diagrams come in many different forms, from simple flow diagram-type images, with maybe only a handful of boxes and one or two connections, through to large and complex diagrams with many boxes, many connections, detailed legends, and lots of annotation and supporting text. What all ToC diagrams have in common is that they are attempting to map out the connections and pathways between an intervention and its outcomes. They all use some form of causal logic to describe what and how impacts might be created by an intervention. They are all intended to explain the 'logic' or 'theory' of the intervention. All but the simplest examples do this by using the boxes, connections, and any text to describe the elements of the intervention, its immediate outputs, longer-term outcomes, and ultimate impacts. Importantly, any key assumptions about how these will be realised are typically included in the diagram.

Let us look at some examples to flesh this out. Figure 3.1 shows an example of a ToC diagram made to show the ToC of a child support grant programme in South Africa. It uses a top-to-bottom layout; on the right-hand side we can see the key categories of 'activities', 'outputs',

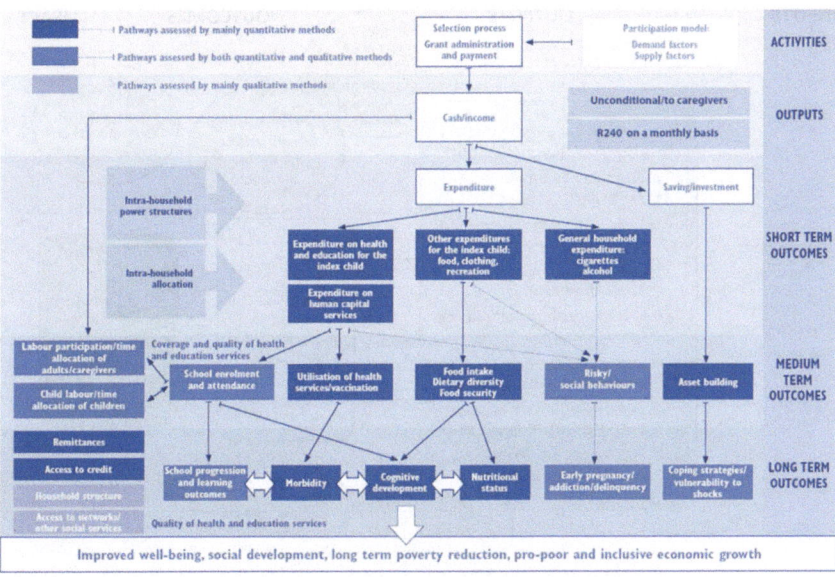

Fig. 3.1 Theory of Change diagram for a child support grant programme in South Africa. Source: DSD, SASSA, and UNICEF (2012)

'short-term outcomes', 'medium-term outcomes', and 'long-term out-comes' (see below for definitions). Within each grouping there are boxes which refer to different types of things. These are connected by different types of arrows. As in many ToC diagrams, shading, additional boxes, and annotations are used to highlight different points. The diagram has many ambiguities about exactly what arrows and boxes can mean or can be, this is normally explained in accompanying text, but you should expect variety and ambiguity in the definitions. We see this is a ToC diagram with a medium level of detail or complexity. There are only a few activities and outputs, but many types of outcomes. In other ToC diagrams, we might see more emphasis on different components of an intervention, that is, more activities, or another common category—'inputs'. The boxes and annotation at the top left make clear this ToC is being used as part of an evaluation, with the reference to the types of methods which will be used to assess different elements.

Because there is such variety in ToC diagrams, it is worth looking at another example. Figure 3.2 shows an example for an education

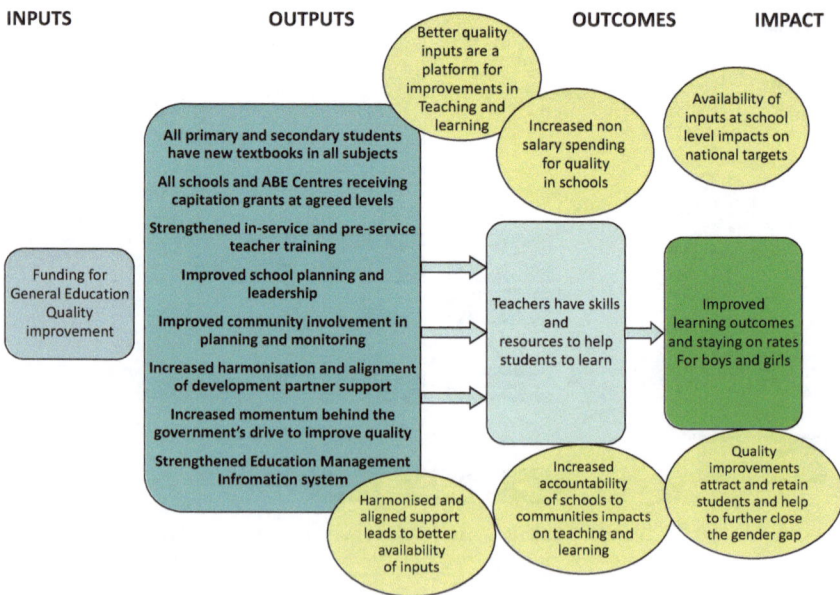

Fig. 3.2 Theory of Change diagram for an education improvement programme in Ethiopia. Source: Vogel and Stephenson (2012)

improvement programme in Ethiopia. This is a simpler diagram and uses a left-to-right layout. There is only one input (the funding for the programme), no activities listed, and then a list of aspirational outputs which are rather abstractly connected to outcomes and impacts. The diagram is likely useful as a quick communication tool but, because of its simplicity, does not really create any additional framing or analytical value.

The differences between these two examples are clear, but they do have some things in common. The most obvious thing is that they both use some basic units or categories to structure the diagram (inputs, outputs, etc.), and they both use a direction of flow, from one side of the diagram to the other (we have seen examples which don't do this, but they are rare). The exact list of the categories differs in many diagrams and are often a source of confusion. We have tended to see the following common categories, which are also discussed in Rogers (2014) (referred to as a 'results chain'):

- **Inputs**: the resources (broadly defined) used or required.
- **Activities**: the actions, events, and undertakings of the intervention.
- **Outputs**: the immediate tangible products of the intervention. These tend to be easy to define and identify, akin to something like deliverables from a project.
- **Outcomes**: the potential short and medium-term effects of an intervention. These might be more difficult to measure and will be less tangible than an output.
- **Impacts**: the long-term effects of an intervention and/or the long-term changes it contributes to.

Both diagrams are also underpinned by a causal logic. Even though they differ in level of detail, they are trying to articulate the cause-and-effect relationships between interventions and outcomes. This is not done at an individual variable, factor, or mechanism level (as in some of the methods in this book) but a more aggregate abstract level. Nonetheless, the causal element is important. Arguably, all the causal assertions in ToC diagrams are based on trying to show what aspects of an intervention are necessary to create the changes they are aiming at. In the first diagram, there is enough detail to see these causal pathways as something like causal mechanisms by which the intervention leads to the long-term goals. In the second diagram however, cause is less clear. One could argue that the outputs are not connected closely enough to inputs and this could mean that potential enablers or barriers to effectiveness of the intervention are not addressed.

Though the clue is in the name, it is important to keep in mind that ToCs are theories, they represent the mental models of the people who constructed them. They are not maps of reality and they may contain gaps and ambiguities, as indeed may the theories they represent. When used in the evaluation of an intervention, one of the core purposes of the evaluation is to test this underlying theory and ask, 'did this intervention have the effects we hoped it would, i.e. is our ToC correct?' This is important to remember because it affects the way we might use a ToC diagram. As a theory, we should see it as a rough guide, one that might be incomplete and foggy in places, but one that will become clearer as we evaluate and plan in more detail. A ToC should ideally be iterated multiple times, being refined each time.

As we stated above, there is a lot of variety in the practice of ToC diagrams, but there is also variety in terminology which can be confusing. There are many phrases that are seemingly used interchangeably with ToC, these include: 'programme theory', 'intervention theory', 'logic mapping', 'logic models', 'results chain', and 'outcome mapping'. We do not want to attempt to define these here, which is a difficult and thankless task. ToC is often used as an umbrella term for a process within which things like outcome mapping or logic mapping might be done, or an intervention/programme theory might be developed which includes ToC diagrams. It is important to take into consideration the fact that these terms are used in different ways. Some researchers and practitioners have tried to define them precisely, but such is the variety in practice and terminology, that we have found we normally cannot rely on these labels to understand what someone is doing, but rather need to look at the work directly.

How Do You Create Theory of Change Diagrams?

Due to the variety of practice in ToC we cannot outline a definitive or detailed step-by-step guide on how to create ToC diagrams, but we can outline some of the broad steps involved. These can be undertaken in a workshop setting with stakeholders, or in small teams of those directly involved with an evaluation or design process. Occasionally, ToC diagrams are developed by individuals, but we would normally advise against this unless there are very clear reasons why this is appropriate in a given context. As with many of the methods in this book, diagrams can be created with pen and paper, post-its, and so on, or with software, though using software greatly privileges the role of the person who is operating it.

Broadly speaking, the following steps are common in the creation of relatively sophisticated ToC diagrams; if you want to create a simple ToC for communication purposes, these steps are probably overkill:

- **Start with the intervention:** in many cases it will be clear what exactly the intervention is, and what inputs and activities make it up, so this may be an easy step. The specific elements of the intervention must be agreed upon and turned into the boxes which fall in the 'inputs' and 'activities' categories of a ToC.
- **Define the long-term impacts:** next, we jump to the other end of the diagram to the long-term impacts the intervention is trying to affect or contribute to. These tend to be similar to the basic aims or goals of the intervention and are normally quite clear from the start. Again, they should be broken down into distinct elements which can be placed in boxes on the diagram.
- **Fill in the gaps:** now comes the first difficult part, starting to fill in the gaps between the elements of the intervention and its long-term goals. You will need to choose which set of categories you want to use, we tend to prefer the set we define above. To fill in the gap, we recommend starting at the short and medium-term outcomes which are closely related to the long-term impacts, defining these, and then defining the outputs of the intervention that lead to them (i.e. starting towards the end, and working backwards).
- **Make it specific and realistic:** as you go, you will likely need to prompt regularly the people involved to be specific and realistic in what they are suggesting. A realistic set of steps which create the pathway or mechanism between intervention and impact should be clearly visible in the diagram. If any step or box feels as if it hides or simplifies away important detail, this should be explored and captured. It can be useful to ask people for specifics by always asking who will be doing something, how much of something will be happening, how that will lead to the next thing, and so on. Another framing which often makes things more realistic is to ask people about what risks they are worried about which might undermine the processes they are describing.

- **Surface the assumptions:** prompting people to be specific and realistic will likely lead to the surfacing of a lot of assumptions of how interventions will work and how people and organisations will react to them. Where assumptions feel particularly important, contested, or uncertain, we recommend capturing them in the diagram, either with boxes and connections or with annotations.
- **Negative or no change:** a powerful way of forcing realism in peoples' thinking is to ask them to also specify 'theories of no change' or 'theories of negative change' in the diagram. In the former, we are looking for the reason why an intervention might have no effect or why a particular thread may not deliver the change we are hoping for. Taking this further, it can be useful to prompt people on potential negative effects and impacts, or unintended consequences of an intervention, and to capture these too.
- **Capture feedbacks and interactions:** it is quite normal for ToC diagrams to start with connections which all flow in one direction, or with 'lanes' of connections in parallel which don't interact with each other. It can be a useful exercise to look for interconnections and interaction between the different pathways you have created. Similarly, asking people for feedbacks and connections which flow in the opposite direction can be a useful prompting technique for building a more nuanced and realistic ToC diagram.
- **Capture disagreement:** areas of disagreement are often key issues which are worth capturing in the diagram and focusing on in an evaluation. Our aim should not be to resolve these immediately but to note them and test them later.
- **Use it, then iterate:** it can be tempting to aim for a perfect or 'finished' diagram straight away. However, this is a very difficult ask. In an evaluation context, it is much more common, and useful, to think of your ToC diagram as a living document, which can and should be refined and updated as an evaluation unfolds. Once you have something usable, get out and start using it, test it, critique it, refine it. We do not discuss in detail here how to use a ToC as part of an evaluation, for guidance on this we recommend starting at https://www. betterevaluation.org/en/resources/theory-change-thinking-practice-stepwise-approach.

The exact details of your ToC diagram process will depend on many factors such as the purpose of the ToC, your preferences, the project needs, your resources, the level of stakeholder engagement, and the size of the role a ToC diagram is to play in a project. There is huge flexibility in the method, you can start with a large stakeholder workshop, or you could create a diagram in a small team and then gather feedback from stakeholders. You may want to use the diagram as more of a communication tool, as opposed to a framing or detailed description of an intervention. In this mode it is common to opt for simpler diagrams which remove some of the more nuanced steps we have described above.

Common Issues and 'Tricks of the Trade'

There are many potential issues you may bump into with ToC diagrams, but two of the biggest are (i) the multiple pressures to create a narrow or overly simple ToC and (ii) the tendency of people (stakeholders, or clients if you have them) to treat a ToC diagram as a product rather than a process. On the first issue, there is an endless list of pressures which push you towards creating a relatively simple ToC diagram. You may not be able to speak to many stakeholders or get input from people with a variety of views. You may need to put the diagram in a report or a slide deck which means it needs to be readable at A4 size or at low resolutions. You may have teams which are responsible for an intervention that insist on keeping a narrow focus on elements close to their control. These are all powerful forces in their own right, but together can make it impossible for you to stop feeling that you must create a very simple ToC diagram. In some cases, it may be appropriate to have a simple ToC diagram, but we believe in most situations it is useful to resist these pressures if we want an effective ToC to frame, structure, and implement an evaluation or design process. You will likely need to get used to deploying the best arguments for resisting these pressures. It may also be helpful to have two versions of a diagram, the 'real one' and a simplified one, which you can use when you cannot resist the pressure to simplify.

On the second issue, it is common to find an over-emphasis on the 'product' of a ToC process, normally the diagram itself. This can come from your client or the team using the ToC, but it may be something you naturally tend towards. We are of the view that we should try to maintain a clear understanding of the development of a ToC diagram as an ongoing process, which we iterate through, and may never really 'finish', but rather have 'pause-points' for. Alongside this, it will likely be useful to cultivate an understanding with clients and stakeholders that the value of ToC diagrams, and the wider approach, is just as much in the process of developing and using it with a wide group of people, as it is with the product of the diagram itself.

To deal with these, and other, issues, the following 'tricks of the trade' may be useful:

- **Always ask 'how, who, how many, why?':** peoples' mental models are almost always vague. Our job becomes helping to translate those vague mental models into specific assumptions and assertions about an intervention. Whenever people want to add a box or a connection to a diagram, we should be asking how that leads to change, who is going to be involved, how many people are involved or how much money is being spent, and why they have confidence that something leads to something else.
- **Question the scale of impacts people describe:** another common theme in peoples' thinking, especially those invested in an intervention or dedicated to making change, is that they overestimate the scale of influence something they are attached to may have, and underestimate the influence of everything else going on, and/or the general inertia around persistent challenges. Where people have got specific ideas down in a diagram, we should follow up by trying to make them more realistic. One way to do this is to focus on scale. Will that series of five workshops really change the direction of an industry? Will that outreach programme really change attitudes if only the usual suspects take part?
- **Use text and annotation around a diagram:** ToC diagrams take numerous forms but many of the best make liberal use of annotations and text around the edges to provide nuance and context. Don't be afraid to do the same, this will help you communicate complex ideas efficiently.
- **Make workshops fun:** ToC diagram categories and their very applied nature can make it feel like a bureaucratic, technical, and dry process to build them. Workshops with stakeholders work best when they are energised, relaxed, and having fun. Try to design your workshop to help foster this environment. Keep them as informal as possible, keep people on their feet, avoid setting the 'rules of the game' too tightly (i.e. don't pester people on getting definitions exactly right, use your and their energy to focus on getting specific and realistic diagrams).

What Are Theory of Change Diagrams Good and Bad At?

Hopefully, through this chapter the strengths of ToC diagrams have become clear. They are excellent at representing stakeholders' mental models of how an intervention leads to outcomes and long-term impacts. They have been very successful and popular in helping people frame and structure the evaluation and design of interventions. They help us surface and make explicit the assumptions which interventions rely on and make their logic more realistic, clear, and shareable. They help us to examine and test this logic and our assumptions by making them visible and facilitating discussion and self-reflection amongst those involved. They help communicate the ideas behind an intervention and can create the appearance, and often the reality, that we have really thought through what we are hoping to do. They can capture important context which may need to be examined as part of an evaluation and by coding or annotating the elements of the ToC can aid in evaluation planning and reveal where gaps in coverage might exist.

However, despite its popularity, the approach does have weaknesses. They can prevent us from taking a whole-system view. The focus on a sole intervention, and not the wider system, can narrow down our perspective quickly, stopping us from seeing wider contexts that are important. They can also reinforce our innate optimism; the inherent focus on the outcomes and impacts we want can blind us to potential unintended consequences, negative impacts, and the underlying inertia in systems we care about. Lastly, the focus on intervention can also exclude people who are not invested or involved in that intervention. Unlike some of the other methods in this book, it might be impossible for them to see their position in a system, or their reality, reflected in a ToC diagram, and so it may become difficult for them to engage.

A Brief History of Theory of Change

Though the ideas and some of the terminology around ToC can be traced back further, the birth of ToC is often attributed to the US-based Aspen Institute and its roundtable event in 1995. The event was organised to discuss the evaluation of community change initiatives. At the time, there was a growing understanding of two fundamental issues: (i) experimental evaluation methods were not appropriate for the large poverty reduction programmes being designed and implemented, and (ii) stakeholders in complex policy programmes were often unclear on how interventions unfolded to create the changes and impacts they sought. The event led to the publishing of a book in which Carol Weiss outlined these ideas and coined the term 'Theory of Change' to refer to the description of all the steps that lead to the long-term goals of interventions (Weiss, 1995). This built on the existing idea of theory-based evaluation and has become a core part of this approach to evaluation.

Since then, the approach has been extremely popular and has been adopted in many different domains, most notably in international development, philanthropy, and the third sector. It is used in all sorts of organisations, from small charities and non-governmental organisations, through to government departments and large businesses, right up to the United Nations. In its origins, the motivations to use the method in a whole-systems and complexity-appropriate manner were present (though the language of systems and complexity were not used), but these have often been lost in its application (see the pressures to create narrow ToCs we describe above), likely due to its sheer popularity and use in many different domains. In turn, and somewhat ironically, there have been many criticisms, often around ToC being too linear and not considering wider contexts (Wilkinson et al., 2021). This is now one of the main sites for innovation in the method, with efforts to make ToC more systems or complexity-appropriate coming from many places (full disclosure, this is an effort that we have contributed to). There have also been efforts to build ToC diagrams that emphasise the actions of specific actors, rather than often agency-free mechanisms (see Van Ongevalle et al., 2014), and serious efforts to reflect on the method and its wider set of challenges, both technical and conceptual (see Davies, 2018).

GETTING STARTED WITH THEORY OF CHANGE MAPPING

Owing to its widespread use, there are many resources on getting started with ToC, and there are likely to be guides or examples relevant to the domains or disciplines you are working in; do look for these. We particularly recommend the following useful resources:

- **www.theoryofchange.org**: run by the Centre for Theory of Change, this site has a wealth of detailed resources and guidance for ToC. It also is the home of the 'TOCO' software for ToC diagrams. We have not used this software ourselves and have had mixed reviews from colleagues that have; it is not free but is worth considering as a software option.

- **Logic mapping hints and tips guide** (Hills, 2010): this guide was prepared for the UK Department of Transport and provides an invaluable, short, and practical guide to the individual steps within a workshop to build a ToC diagram. Note, the use of different terminology!
- **Review of ToC in international development (Vogel, 2012):** this review, prepared for the UK Department of International Development, is much longer than the one prepared for the UK Department of Transport, providing similar detailed guidance but also much wider conceptual discussion. International development is one of the fields in which ToC has been most used, so there is a lot of material to be learnt from here.
- **UNICEF Methodological brief** (Rogers, 2014): this short guide provides another perspective and is a useful complement to the two reviews above.

Hopefully, you now have most of what you need to get started using ToC diagrams yourself. We would strongly recommend looking at some of the resources above, looking for examples in similar domains and contexts to yours, and reaching out to practitioners and researchers using ToC in similar ways or areas to you. Learning from experienced ToC users is perhaps the quickest and most powerful way to learn the near-tacit skills required which is hard for us to get across in a book chapter. It tends to be easy to find examples of ToC diagrams you like, and to draw inspiration from them (you should do this!), but it is much harder to find processes that you like and that you want to emulate. Speaking to experienced ToC users will help you do this. Good luck!

REFERENCES

Davies, R. (2018). Representing theories of change: Technical challenges with evaluation consequences. *CEDIL Inception Paper 15.* https://cedilprogramme. org/publications/representing-theories-of-change-technical-challenges-with-evaluation-consequences/

DSD, SASSA and UNICEF. (2012). *The South African child support grant impact assessment: Evidence from a survey of children, adolescents and their households.* UNICEF South Africa.

Hills, D. (2010). *Logic mapping: hints and tips guide*. Department of Transport. https://www.gov.uk/government/publications/logic-mapping-hints-and-tips-guide

Rogers, P. (2014). Theory of Change. Methodological briefs – Impact evaluation No. 2 UNICEF. https://www.unicef-irc.org/publications/747-theory-of-change-methodological-briefs-impact-evaluation-no-2.html

Van Ongevalle, J., Huyse, H., & Van Petegem, P. (2014). Dealing with complexity through actor-focused planning, monitoring and evaluation (PME). *Evaluation, 20*(4), 447–466.

Vogel, I. (2012). Review of the use of 'Theory of Change' in international development. Department for International Development. https://www.gov.uk/government/news/dfid-research-review-of-the-use-of-theory-of-change-in-international-development.

Vogel, I., & Stephenson, Z. (2012). Appendix 3: Examples of theories of change. In *Department for International Development*. https://assets.publishing.service.gov.uk/media/57a08a66ed915d622c000703/Appendix_3_ToC_Examples.pdf

Weiss, C. (1995). Nothing as practical as good theory: Exploring theory-based evaluation for comprehensive community initiatives for children and families. In J. P. Connell (Ed.), *New approaches to evaluating community initiatives: Concepts, methods, and contexts* (pp. 65–92). Aspen Institute.

Wilkinson, H., Hills, D., Penn, A., & Barbrook-Johnson, P. (2021). Building a system-based theory of change using participatory systems mapping. *Evaluation, 27*(1), 80–101. https://doi.org/10.1177/1356389020980493

CHAPTER 4

Causal Loop Diagrams

Abstract This chapter introduces Causal Loop Diagrams. We explore what exactly Causal Loop Diagrams are, describe how you can use them, take a step back to consider common issues and 'tricks of the trade', as well as present a brief history of the development of the method. This chapter can be viewed as a companion to Chap. 8 on System Dynamics; these two methods are closely related. Causal Loop Diagrams emerged from Systems Dynamics practice, and though it is a systems mapping method in its own right now, it is still often used as a stepping-stone to the development of System Dynamics models.

Keyword Causal Loop Diagram

Let's get the elephant in the room out of the way. The obesity system map (Government Office for Science, 2007), which you have likely seen, was produced for the UK government back in the mid-2000s and is a Causal Loop Diagram (CLD). It was, and still is, one of the most high-profile pieces of systems mapping work ever done and has been very successful and influential by most measures. However, it is also often derided or held up as an example of a 'horrendogram' (Penn, 2018); indicative of a form of visualisation which renders problems intractable and overwhelming.

Even without the obesity example, CLDs are a no-brainer to include in this book. They are a well-used approach to systems mapping, with roots

© The Author(s) 2022   47
P. Barbrook-Johnson, A. S. Penn, *Systems Mapping*,
https://doi.org/10.1007/978-3-031-01919-7_4

going back to the 1970s at least, and a strong connection to System Dynamics. The method sits roughly in the middle of the qualitative-quantitative spectrum of systems mapping methods, perhaps more towards the qualitative side, but hints at dynamics in a system. It focuses on feedbacks as a key component and organising structure for complex systems, but also brings the ability to include all sorts of concepts and variables. This means it can facilitate the production of focused maps, which strongly hint at the dynamics of systems, but also much larger inclusive ones, which allow us to see the big picture and the interconnected nature of systems. CLDs also have one of the most consistent and appealing visual styles of any of the qualitative methods in this book. Arguably, this started with the influential 'Foresight' work (done in the UK), which showed what proper resourcing for design could do for developing high-quality visualisations. This tradition has been extended with contemporary software tools, such as Kumu, which place a high value on slick and appealing visualisations. Finally, it is worth noting, CLDs are a natural stepping-stone to simulation methods such as System Dynamics because of their focus on feedbacks and strict use of variables for nodes in the map.

The rest of this chapter will follow the same structure as the other methods' chapters in this book. First, we describe exactly what a CLD is. Next, we describe how the method is 'done'. Then we consider common issues and some 'tricks of the trade' for using the method. Next, we reflect on what the method is good and bad at, before exploring a brief history of its development. We finish with some useful resources and tips for getting started with the method yourself.

What Is a Causal Loop Diagram?

CLDs represent a system in three basic elements: boxes, connections, and feedback loops. The boxes, or nodes, represent variables in the system; these can be anything as long it makes sense to think of them going up or down over some scale. The connections, or edges, represent causal influence, from one node to the other; either positive (i.e. they increase or decrease together) or negative (i.e. they change in opposite directions, if one goes up, the other goes down, and vice versa). So far, fairly similar to Fuzzy Cognitive Mapping or Participatory Systems Mapping. The third element is what makes CLDs more unique. The maps always show and focus on feedback loops, both in the construction of the map and in its visualisation. Loops are made conspicuous by the use of curved arrows to create circles. The loops are also sometimes colour coded to highlight them or are annotated with small arrows and '+' or '-' symbols to highlight if they are reinforcing (positive) or balancing (negative) feedback loops.

The feedback loops are usually focused around a 'core system engine', which is a set of nodes that are the core of the system. These are often visualised more prominently than other nodes in the map. The feedbacks strongly hint at dynamics in the system. It is common to use the maps to think at a slightly higher level than individual nodes and edges, bringing together the handful of feedbacks in the system to think about how these might play out together. The focus on feedbacks means the CLDs are a relatively disciplined way of looking at a system, which places the existence and effects of feedbacks at its core. It is common to see both simple (perhaps five or ten nodes) and more complex CLDs (dozens and dozens of nodes).

Let's look at some examples. Figure 4.1 shows an example of a simple CLD. We can see how nodes and edges are connected with positive and negative connections, and how the feedback loops of 'revenue generation', 'organizational legitimacy', and 'social action' are emphasised through their positioning, and additional annotation. The 'R1', 'R2', and 'B1' refer to reinforcing and balancing loops.

The next example, shown in Fig. 4.2, is the well-known and much larger CLD of the obesity system in the UK. Here, nodes are in boxes and colour coded by theme, and the connections are shown with solid lines (positive connection) and dotted lines (negative connection). The core engine of 'energy balance' and its feedbacks are highlighted in the centre.

CLDs are used in a variety of ways. Most fundamentally, as with many methods in this book, they are a way of surfacing, visualising, and exploring mental models. Exploration of the full map is done in many ways, but qualitative analysis of the core engine and feedback loops is a common focus. CLDs are often a precursor to System Dynamics models (Chapter 8); used to begin the modelling process in an intuitive way, before the conversion to stock and flow diagrams and differential equations.

The maps can be built from all types of information, and in participatory modes, and the mix between these is fairly even in the literature. There is a lot of variety in the ways they are built and the exact purpose they are put to. However, there is less variety in what is actually constructed; the use of clear variables, arrows for influence, and focus on feedbacks is consistent. There is a lot of quite prescriptive guidance available,

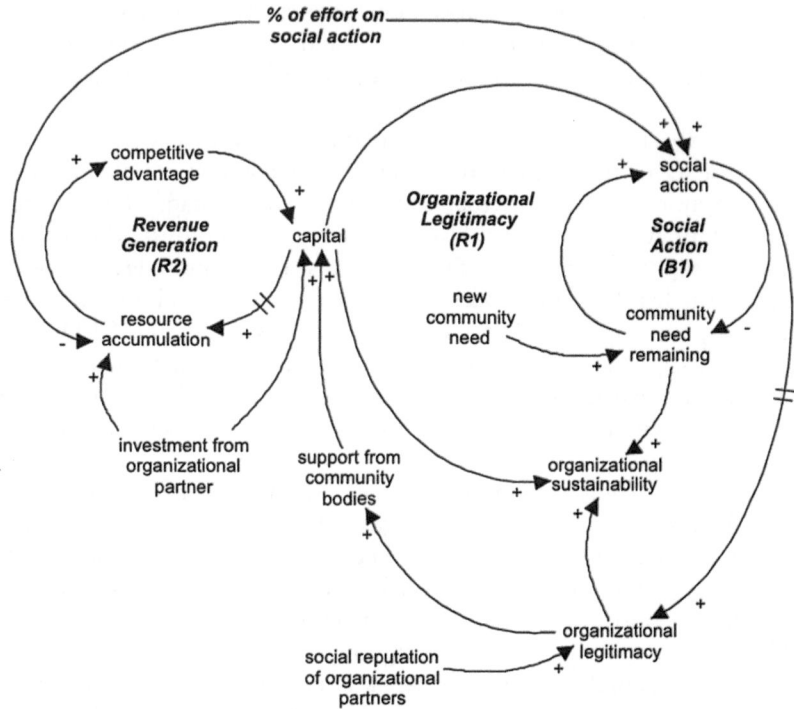

Fig. 4.1 A Causal Loop Diagram of the tensions between business activity and social action in a social enterprise. Note, double bar (//) symbols indicate a time delay. Three feedback loops are emphasised (B1, R1, and R2). Source: Moizer and Tracey (2010)

which helps to maintain this consistency. However, it is worth noting, there does appear to be a divide between those who use CLDs as a stand-alone method and those that use them as a stepping-stone to System Dynamics. The practice of each group does tend to have subtle differences, with the former being more inclusive and flexible in their use of the method, and the latter following clearer rules about how to develop a map. There is a little variety in the terminology used to describe CLDs; they are often referred to as 'influence diagrams', or simply 'system maps'. There is also a small group of papers which use the term 'Participatory

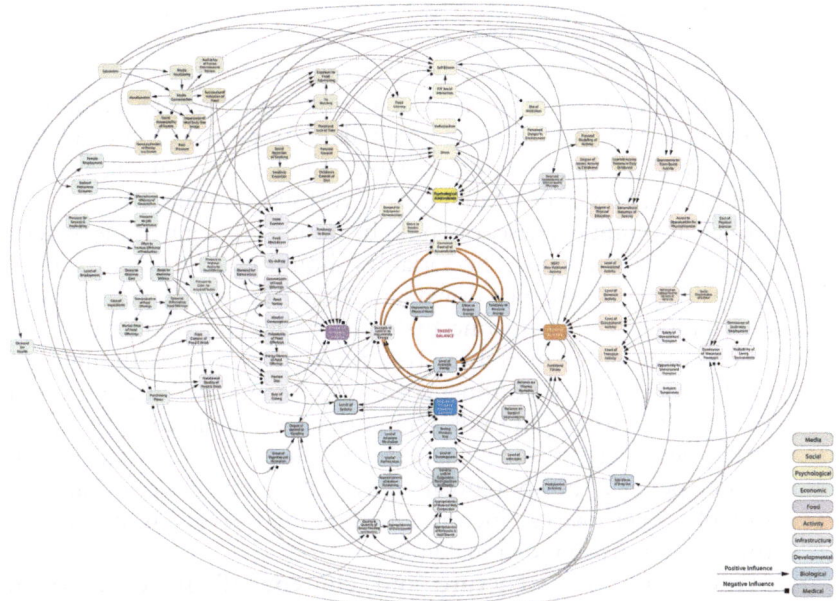

Fig. 4.2 A Causal Loop Diagram of the obesity system in the UK. Source: Produced by ShiftN for Government Office for Science (2007). The core feedback loops are emphasised in the centre

Systems Mapping' to refer to a participatory CLD approach (e.g. Lopes & Videira, 2016).

How Do You Create Causal Loop Diagrams?

The process of building and using a CLD is typically iterative. They can be developed in a participatory workshop mode or based on data (typically qualitative) and evidence. However it is done, the process tends to include the following steps:

1. **Collect data and/or evidence:** first you need to know what you are going to build your map from. Maps can be built in a participatory mode, in which case, discussions during workshops will be your data. However, because of the strong discipline needed to focus on

a system engine and feedbacks, decisions are often made by the modeller/researcher. When built from data or evidence, these need to be collected or compiled before mapping can start.

2. **Screen data for variables and connections, create catalogue:** once you have your underpinning data or evidence, you need to screen and extract potential variables and connections from them, and build a catalogue of them. In a workshop setting, this could be a brainstorming exercise, or could be done at the same time as actually drawing the map. The variables should be expressed clearly as things that can go up or down and should be precise. Try to avoid developing 'container concepts' such as 'Technology' which can mean many different things. If a connection feels as though it is unclear or complex, consider if adding a second will make it easier. For example, we might think a 'price of good' variable has an unclear influence on a 'revenue' variable (i.e. higher prices equal more money per sale, but potentially fewer sales), but if we add another variable, 'sales', we can incorporate this effect separately and remove the uncertainty in individual links.

3. **Hypothesise the structure and content of the core system engine:** this is the first CLD-unique stage. Assuming you already have a strong sense of the purpose of the mapping project, and you understand users' interests, this is the stage at which you start to focus in on what might be the 'core system engine'. There is no simple way to decide this. It may be that there is one defining variable at the centre, and that is enough; or you may have a handful of key variables which are of most importance to people in the system, or which interact in ways which drive system behaviour. Alternatively, you may have an idea or narrative about what sort of dynamical behaviour is at the heart of the system. The engine will be the focus of the map as a whole and will be one of the main focuses for exploring feedbacks. At this stage, you will need to identify which variables are included in the engine and begin to develop ideas about how they might be connected. This might be as far as you hope to get in a first workshop, before stopping and starting the next stage in a desk-based research mode. This stage can be difficult, it relies on craft and judgement more than objective fact or technical skill. You may find it useful to refer to the 'system archetypes' developed by Kim (2000), which outline common and intuitive types of core system engines; they are: 'drifting goals', 'escalation', 'fixes that fail',

'growth and under-investment', 'limits to success', 'shifting the burden', 'success to the successful', and 'tragedy of the commons'.

4. **Develop first diagram with core system engine and twenty nodes:** now you are ready to build the first diagram of the core system engine and the variables immediately around it. This will be a stage with many iterations and repeats as your ideas crystallise. Once you have the engine down, you can start to build in a similar way to other methods, either brainstorming (or extracting from your catalogue) the top twenty variables that are influenced by, or influence, the core, and connecting them; or by more organically building one variable at a time. There are some excellent and detailed guidance for this stage at www.thesystemsthinker.com, including how to deal with ambiguities, including delays in influence, and naming feedback loops. Twenty nodes seem to be a useful limit here, beyond this the core becomes too unwieldy.

5. **Verify and amend:** this stage is arguably the most important, but the most underestimated. Once you have a first version of the core system engine and surrounding nodes, you need to expose it to as much critique, feedback, and commentary as possible. This may be with stakeholders who you are working with, with clients or users, or with other researchers and modellers. You should amend the core system engine considering this feedback and your aims or create different versions if necessary.

6. **Expand model as needed:** once the core feels more settled, you can expand the map if needed. It may be that you want a simple map, and the core you already have is enough, in which case you can skip this stage. Or, you may want to build a much larger map, in which case you should expand and then iterate through another 'verify and amend' stage.

7. **Analysis and use:** once you have the map, there are several ways it can be used and analysed. Purely by its existence it makes clear your, or your stakeholders, mental models. You might want to do a thematic analysis of it, either in written form or in a workshop setting with stakeholders. You may want to explore it qualitatively, that is, investigate how the various feedbacks in the map might play out and ask questions such as 'is the system in a stable lock-in or might there be reinforcing feedbacks which cause runaway change?'. It is common to create mock-up plots of variables through time to explore these. Finally, if you are using the map as a starting phase for a

System Dynamics project, you are now ready to start the process of converting it; there is no simple recipe to follow for this. At this stage, it is also common to finalise the visualisation of the map; many CLD projects put a lot of value in a well-designed and appealing final map. You may want to consider asking a designer to help you with this.

Commons Issues and 'Tricks of the Trade'

Most of the issues that are more specific to CLDs tend to revolve around the use of feedbacks. Constructing feedback loops requires a different way of thinking, and perhaps a little more care than other parts of the map, or other types of maps where we tend to work solely in variables and connections. It brings an additional layer of easy mistakes to make. Fundamentally, this is because we are building a group of important and interacting variables and links together. This means we need to hold more information in our minds at once, and imprecisions can creep in more easily. Guidance on CLDs often provide tips such as labelling loops or making the goals of balancing loops clear (i.e. including a variable which expresses a goal and linking it to a loop). These are specifically aimed at helping loop construction go well.

When it comes to interpreting and using CLDs, it is easy for the feedback loops to be misunderstood or skipped over. Many people will find it hard to link together multiple loops and think about how these may interact through time. It can be useful to create mock-up plots of variables through time to explore how the system might behave. Others, rather than finding loops difficult, may simply not realise their significance in a CLD and ignore them. This is one of the easiest ways to miss one of the key values of CLDs. Even when loops are noticed, understood, and discussed, it can be difficult to turn this dynamical view of a system into something usable or 'actionable'. The map will give us the language and understanding on an individual level, but problem-solving cultures in organisations can get in the way if they revolve around instrumental and 'button-pushing' mindsets. This is an issue with most systems mapping methods, but it really comes to the fore with CLDs because of the combination of accessibility and exploration of dynamics. The system archetypes described by Kim, for example, are accompanied by suggested ways to reframe problems or redesign situations to help change dynamics, but this certainly requires decision-makers taking a systems view. One more straightforward action might be identifying simple positive/reinforcing, or negative/balancing, feedbacks, which might be encouraging or supressing change that we do or don't want. Disrupting a 'vicious cycle' or encouraging a 'virtuous' one is probably the most intuitive way in which we can think about interaction with system dynamics.

There a few 'tricks of the trade' to help us deal with these and other issues, these include:

- **Tune the complexity of a map to the visual literacy of your audience:** while we often talk about bringing users, clients, and stakeholders into our thinking on purpose and design of a systems mapping project, we don't normally talk about their visual literacy. CLDs have good tradition of taking real care over visualisation, and part of this includes thinking about the visual literacy of your audience. If you think your audience will be put off by large maps, or be confused by multiple interacting feedback loops, think about how you can tune your map to this, start small with the engine, and build up. The point should not be to just produce a very simple map, though you may end up doing this, but to also think carefully about how you might frame and introduce it in ways which allow you to keep as much complexity as possible (e.g. introduce it in steps, or colour code and position nodes by themes).
- **Allocate half of your resources to design and communication:** it is natural to focus our energy on creating a map. However, when we are using maps with subtleties beyond face value and that need to be created with a lot of control by researchers (i.e. like CLDs), we really need to invest in our final visualisations and communication of maps to users and stakeholders. Otherwise, we will have created a brilliant CLD which no-one is using or looking at. This takes time and effort, we would suggest thinking about using half of your time or resources on this; it can include a range of activities, from smart visuals, web-based interactive maps, to events to disseminate and use the map.

WHAT ARE CAUSAL LOOP DIAGRAMS GOOD AND BAD AT?

CLDs' real strength comes from producing a system map that is often visually appealing, that hints strongly at a system's dynamics, but that is still flexible, inclusive, and relatively easy to use. CLDs can make use of all sorts of information, can be big or small, and offer a clear stepping-stone to producing a more formal System Dynamics models.

On the downside, CLDs can be restrictive with their focus on feedbacks and do tend to put a lot of power in the hands of the researcher, especially around the creation of feedback loops. If feedback loops are not present, or important, in your system, then the method will be less useful. Being in the middle of the quant-qual spectrum means CLDs will also not give you any quantitative analysis. Relatedly, without any quantitative analysis, it can be difficult to meaningfully understand how multiple feedback loops will interact.

A Brief History of Causal Loop Diagrams

CLDs' early history is intertwined with that of System Dynamics. In the early work of Jay Forrester introducing System Dynamics, there is no mention or use of CLDs, though he does use flow diagrams as a mid-point between a verbal description and a formal model of a system. CLDs appear to have emerged later as a way of communicating the design of a System Dynamics model. Goodman's (1975) 'Study notes on System Dynamics' is one of the earliest texts to formally outline CLDs and does suggest they can be used as a tool for model conceptualisation (i.e. in the design stage) too. Simple CLDs were also used in the famous Limits to Growth report (Meadows et al., 1972), which is likely to have raised awareness of them.

In the earliest works, CLDs were primarily used to communicate the design of a System Dynamics model. However, their use has spread beyond this to such an extent that we now think of them as a method in their own right. Moreover, when used in conjunction with System Dynamics, it is more common to use them as a stepping-stone towards a model, rather than as a tool to help communicate them after they have been built.

The 'rules of the game' for how to construct CLDs, and what is included in them, have remained relatively stable through the years. There have been debates around relatively minor issues, such as whether links should be labelled with '+' and '-' symbols or the letters 's' and 'o' (to represent 'same direction' and 'opposite direction'). What has clearly changed is the prominence they are given as a method in their own right, the ways we might build them (i.e. not just from a modeller's design, but from data, evidence, and participatory processes), and the ways we might use and analyse them (i.e. with exploration of the perceived behaviour and interaction of loops).

GETTING STARTED WITH CAUSAL LOOP DIAGRAMS

CLDs are one of the methods which is relatively easy to get started with. If you just want to have a play and get a feel for the method, we recommend just diving-in and building one, either on pen and paper or with specialist software. There are numerous guides available which will be useful to have by your side. The process of implementing the tips on your map will quickly allow you to understand how the focus of a CLD is different to other types of system maps.

More generally, there are range of resources we would recommend for helping you go deeper into CLDs, including:

- www.thesystemsthinker.com: this is a website version of the original 'The Systems Thinker' publication which has run since the early 1980s. There are hundreds of short and accessible articles on all sorts of topics, including dozens for CLDs. We recommend using the search function to find CLD articles, as there is not a stand-alone category for browsing.
- **System archetypes (Kim, 2000):** this suite of guides is invaluable for getting into the most important details of CLD. There is a wealth of information in each one, but the eight archetypes, and the discussion around them really gives a sense of what the core engine of a system might look like.
- **Group model building textbook (Vennix, 1996):** though this book is focused on System Dynamics as a whole, it does consider 'qualitative System Dynamics' which includes CLDs, specifically, and is a foundational text for the participatory development of these types of models.
- **Obesity and land use futures examples (Government Office for Science, 2007, 2010:** finally, it can be useful to see CLDs 'in action'. For this we would highly recommend reading two examples of their use in the Foresight work we mentioned above; one on the famous obesity map, and one on land use futures. You can avoid the long-form reports if you want and focus on the separate publications specifically about the CLDs work that was done.

You can build a CLD with pen and paper, or on using general-purpose diagramming software, but it will be quicker, and possibly more visually appealing, to use some purpose-built software, such as:

- **Kumu:** this increasingly popular software provides an excellent web-based interface for building and analysing system maps. Though it is flexible, it broadly uses the underlying logic of CLDs and so is well-suited to them. You can build 'public' maps for free, but there are fees to use the software to create private maps.
- **System Dynamics software (see Chap. 8)**: most System Dynamics software has the capability to create CLDs, for example, iThink/Stella, Powersim Studio, and Vensim.

Hopefully, you now have a strong sense of what CLDs are and how you can use them. Take advantage of the fact that this is one of the easier methods to get started with; have a play around with some software today, or read one of the guides and dive in. Be warned though, you can get hooked easily, and before you know it you will be looking at everything through the lens of feedbacks and core system engines! As a method that is easy to start with, it is possible to go quite far without reading some of the more advanced guides and textbooks; we would advise against this if possible. It can really improve the quality of your mapping work if you check in with these resources regularly; much of their advice will only really sink in once you are up and running. Good luck!

REFERENCES

Goodman, M. R. (1975). *Study notes on system dynamics*. Pegasus Communications.

Government Office for Science. (2007). *Reducing obesity: Obesity system map*. https://www.gov.uk/government/publications/reducing-obesity-obesity-system-map

Government Office for Science. (2010). *Land use futures: Making the most of land in the 21st century*. https://www.gov.uk/government/publications/land-use-futures-making-the-most-of-land-in-the-21st-century

Kim, D. (2000). *Systems Archetypes I*. Toolbox reprint series. Pegasus Communications. https://thesystemsthinker.com/systems-archetypes-i-diagnosing-systemic-issues-and-designing-interventions/

Lopes, R., & Videira, N. (2016). Conceptualizing stakeholders' perceptions of ecosystem services: A participatory systems mapping approach. *Environmental and Climate Technologies, 16*(1), 36–53. https://doi.org/10.1515/rtuect-2015-0011

Meadows, D. H., Meadows, D. L., Randers, J., & Behrens, W. W., III. (1972). *Limits to growth. A report for the Club of Rome's project on the predicament of mankind*. Universe Books.

Moizer, J., & Tracey, P. (2010). Strategy making in social enterprise: The role of resource allocation and its effects on organizational sustainability. *Systems Research and Behavioral Science, 27*, 252–266. https://doi.org/10.1002/sres.1006

Penn, A. S. (2018). Moving from overwhelming to actionable complexity in population health policy: Can a life help? *Artificial Life, 24*(3), 218–219. https://doi.org/10.1162/artl_e_00265

CHAPTER 5

Participatory Systems Mapping

Abstract This chapter introduces Participatory Systems Mapping, a method for building and analysing causal system models in groups, developed by us. The method uses tools from network analysis and focuses on chains of causal connections to develop meaningful and actionable insights with stakeholders. This chapter describes in detail what it is and how to use it, considers what it is good and bad at, as well as describes some of the history of its development. We also point to resources and tips for getting started with the method yourself.

Keywords Participatory Systems Mapping • Complexity • Policy • Stakeholders • Network analysis

This is probably the easiest yet oddest chapter for us to write in this book. Our experiences over the last ten years, as we have developed Participatory Systems Mapping (PSM), are one of the core motivations for this book. We have experienced all sorts of reactions when using PSM, from the positive and enthusiastic, through to negative and harsh criticism. Sometimes people have seemed to think we have invented a whole new way of knowing and seeing systems; others have dismissed us as pedalling little more than repackaged consultancy services. These views are both wrong, but in different ways, yet somehow for the same underlying reasons. Underpinning both good and bad reactions, we believe are fundamental misunderstandings of the aims, methods, and value of systems mapping, and an

© The Author(s) 2022
P. Barbrook-Johnson, A. S. Penn, *Systems Mapping*,
https://doi.org/10.1007/978-3-031-01919-7_5

61

underappreciation of the variety of ideas and practice. As part of our wider commitment to, and belief in, using systems mapping approaches, we felt it was worthwhile shining more light on these methods. In turn, more selfishly, we hoped this would improve our own knowledge and practice, and help us advocate better for PSM!

PSM has evolved a lot over the last ten years, both in simple terms—what we call it and how we frame it—and in how maps are built and how they are analysed. We have tried to find the spaces between what other methods do well, but also to include the ideas and learning from the many people we have worked with using PSM. We have also tried to stay true to our underlying philosophy of a participatory and humble approach to interacting with and managing complex adaptive systems, and to our core idea for PSM - to turn potentially overwhelming complexity (in systems, and in our models of them), into something more actionable, practical, and usable.

We could write reams and reams about PSM; indeed we have in many different places, but here we will attempt to be disciplined and stick to explaining as simply as we can: what it is and how you can use it, common issues and tricks of the trade, what it is good and bad at, and a brief history of how it has evolved. Finally, we will point to some resources for getting started yourself.

What Is Participatory Systems Mapping?

PSM is the process of building, analysing, and using PSM maps. The maps are causal models of a system, represented by a network of factors and their causal relations. They are almost always annotated and layered with information about the factors and their connections. Technically speaking, they are directed cyclic graphs, that is, the connections between factors have arrows and there can be feedback loops in the network. The maps are built by stakeholders, typically through a series of workshops and meetings, and the participatory nature of their development is paramount. So too is the approach to analysis, which uses information from stakeholders, the principles of network analysis, and looks at the 'flow' and chains of causal relationships (which we often refer to as 'causal flow') to create submaps focused on exploring specific questions or purposes, again in a highly participatory and iterative manner.

Let's look at an example to get a feel for the basics of what makes a map. Figure 5.1 shows both a full map, which is too large to read on the

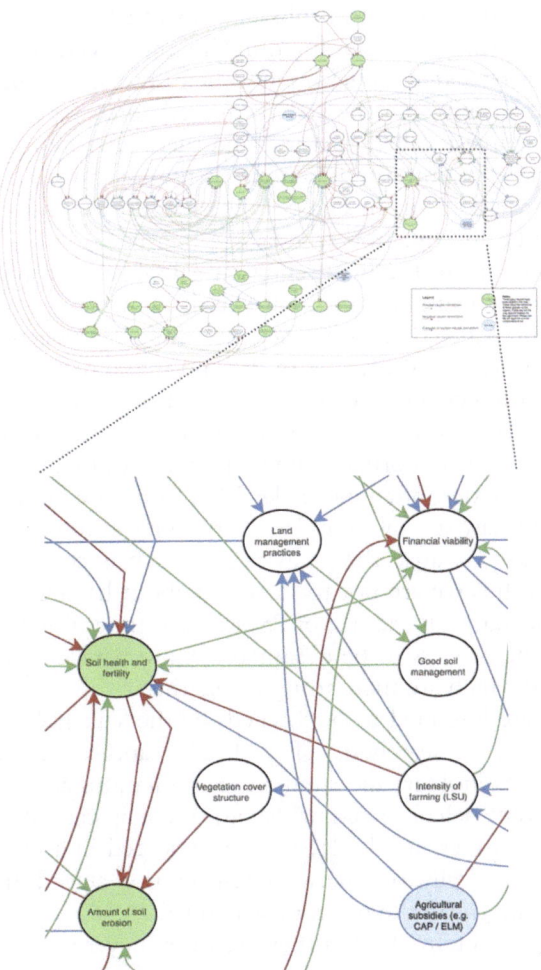

Fig. 5.1 Participatory System Map of the water and agricultural system in a river catchment in north-east England. Green nodes are system functions, blue nodes are policies, green arrows are positive causal connections, red arrows are negative causal connections, and blue arrows are complex or unclear causal connections. Source: Authors' creation based on Bromwich et al. (2020)

printed page, and a zoomed-in subsection, so that we can see the detail. Nodes can be from any relevant domain; they need not be explicitly quantifiable or have data underpinning them, but should be expressed as variables, that is, things in the system that can increase or decrease. The nodes in the map are called 'factors', and there are often special types of factors such as outcomes or functions of the system we care about, or interventions we control. In Fig. 5.1, there are intervention-type factors coloured in blue and function-type factors in green.

The connections in the map represent causal relationships. These can either be positive causal connections (i.e. if A increases, or decreases, B changes in the same direction), negative causal connections (i.e. if A increases, or decreases, B changes in the opposite direction), or uncertain or complex connections (i.e. if causal relationships depend on other factors or contexts, or if the relationship is strongly nonlinear).

In PSM maps we are normally aiming to build reasonably large networks with at least 20 nodes, often more like 50 or 100. The number of connections is then usually several times larger, running into several hundred on some occasions. Because the analysis approach is centred on creating submaps, and thus we are not too worried about having large maps, the only hard constraint on the size of PSM map is time. Building large maps may require the process to be designed to have parallel streams creating different maps which are then brought together.

The maps are intended to be 'owned' by the stakeholders who create them, rather than researchers. They should capture all the complexity important to stakeholders and should use annotations and labels to represent any different beliefs. The final layout of a map can take many forms; sometimes it can recreate the layout that emerged in a workshop, or be rearranged thematically, or a network visualisation algorithm can be used to highlight patterns in the network structure.

Once a map is built (though they are often never really 'finished', we can always add more), we can analyse it. At the core of the approach to the analysis is the idea of creating submaps, that is, subsections of the full map, which we can use to focus in on particular questions or issues. The submaps are intended to allow us to get a handle on the map; they offer us a way in, to what can be an overwhelmingly large diagram. The question now becomes, how do we pick where to extract a submap and what do we include and exclude from it?

The starting point for a submap can be defined either by 'stakeholder-suggested' factors, that is, what factor(s) stakeholders have told us are important, that represent their suggested or current interventions, or that they think are vulnerable to change. Or we can start with 'system-suggested' factors, factors that different types of network analysis (e.g. centrality measures) tell us might have interesting properties in the network. For example, we might have factor with many connections, or a factor which bridges different parts of the map.

Once we have a starting point, the analysis uses one (or a combination) of the following 'rules' to generate a submap: one, two, or three steps 'upstream' or 'downstream' (i.e. following the arrow directions) of the starting factor; ego networks of the starting factor (i.e. any nodes and edges connected to it and the edges between them); or paths between multiple factors of interest (i.e. following the arrows from one factor to the next until we reach the other factor of interest). Figure 5.2 shows how these are defined visually. These rules for generating submaps can themselves also be combined using unions or intersections (i.e. showing multiple submaps together or showing the nodes and edges that are in multiple submaps). We might do this if we want to look both up and downstream from a node of interest, or if we want to see where the ego networks of different nodes overlap. These various approaches to analysis are summarised in Table 5.1.

As the example above and description of analysis shows, the subjective information we collect on factors (i.e. what is important to stakeholders,

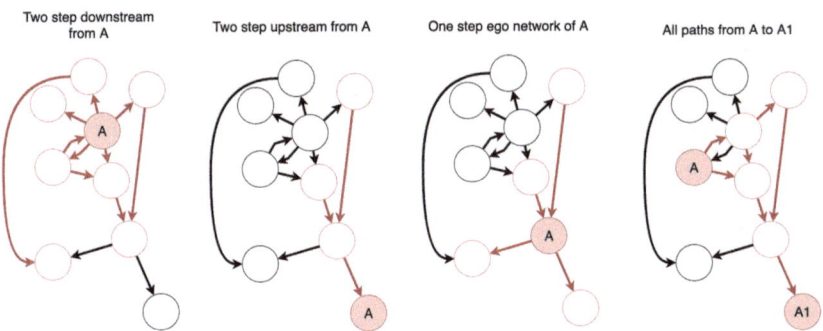

Fig. 5.2 Ways to generate a submap from a starting point. In each network, a submap is created starting from the node A using the mode annotated above each network. Nodes and edges in red are those that will be included in the submap, those in black will be removed/hidden. Source: Authors' creation

Table 5.1 Analysis starting points for Participatory Systems Mapping. Source: Authors' creation

Way to start	Starting point options	How to build	Interpretation
Stakeholder-suggested factors	Intervention or controllable factors	Downstream nodes and edges	What is the intervention or controllable factor affecting? Are there unexpected indirect effects?
		For multiple nodes, create a union or intersection of multiple downstream submaps	How are multiple interventions complementing or clashing with each other?
		Paths between intervention nodes and outcome nodes	What does the intervention rely on to achieve its goals? What wider context might affect it?
	Important or outcome factors	Upstream nodes and edges	What is influencing the thing we care about? Is it vulnerable or supported?
		For multiple nodes create a union or intersection of multiple upstream submaps or ego networks.	What synergies or trade-offs might there be between achieving the things we care about?
		Ego networks	What is influencing the thing we care about, and what does it influence?
		Paths between intervention nodes and outcome nodes	How are interventions reaching the thing we care about?
	Vulnerable to change factors	Up and/or downstream nodes and edges	What might mitigate or drive change in this factor? What impact might change have?
		For multiple nodes create a union or intersection of multiple downstream submaps.	How might changes or risks interact and compound each other?

System-suggested factors		
Influential (i.e. many outgoing connections)	Downstream nodes and edges	What is this influential thing affecting? Is it a lever or a vulnerability?
Central to the map (i.e. well-connected, or bridging parts of the map)	Downstream and/or upstream nodes and edges	What is influencing this central factor? What influence does it have? Is it a bridge or bottleneck?
	Ego networks	What is this central factor close to?
Influenced (i.e. many incoming connections)	Upstream nodes and edges	What is influencing this highly influenced factor? Is it being pushed in one or many directions?
Unusual network property (i.e. low number of connections, but bridges two part of the map)	Any of the above	Does this factor play an important but counter-intuitive role in the system?

what is vulnerable, or controllable) is incorporated into analysis. When combined with network analysis this provides different insights. For example, an influential (high out-degree) factor, which impacts many important functions, is obviously significant. However, if it is vulnerable to change or controlled by an external actor, it may be a vulnerability. Whereas if it is controllable, it may be an opportunity to make change, a so-called system lever. Different types of information can be collected depending on what is relevant to the system and stakeholders. Analysis is also often combined sequentially, with one submap generating questions that lead to the creation of another. In practice, the process of generating submaps is a creative, iterative, and exploratory exercise, ideally done with stakeholders. It can be modified and recombined in numerous ways to address the questions that matter to participants.

There is a reasonable amount of variety in how PSM is used, but this normally stems from the needs and purpose of any given project, rather than different perspectives on how to use the method. Examples of variety include (i) different types of information collected about nodes and edges, such as edge weights, different classes of nodes, or more detail about complex causal relationships; (ii) differences in how a PSM process is designed, with almost infinite options of how to organise sequences of workshops, meetings, and interviews; or (iii) differences in how a full map is connected to other models or forms of knowledge (e.g. using a left-to-right inputs-to-outputs-type layout to make it resemble a Theory of Change map). In terms of terminology, there is not too much variation. PSM maps are not known by another name, but the phrase 'Participatory Systems Mapping' is quite generic, so we have seen it used in a more high-level way to refer to different methods such as Causal Loop Diagrams or bespoke system mapping efforts that emphasise participation.

How Do You Do Participatory Systems Mapping?

We have previously written detailed guidance on how to run workshops (see Barbrook-Johnson & Penn, 2021; Penn & Barbrook-Johnson, 2019) and how to design and conduct a whole PSM project (see Penn & Barbrook-Johnson, 2022). Here, we will attempt to outline the entirety of the process, in a succinct form, and suggest you check our fuller writing on these topics for more detail. The exact nature of your process will depend on the purpose of your project, and you should be creative and flexible in designing a process, while always erring on the side of doing more participation, rather than less.

We would expect to see the following basic stages in almost all PSM processes (note, we will assume that all are done with stakeholders right from the start):

- **Deciding on aim of project:** There are multiple purposes which a PSM process could be used for, for example, solving a problem or asking a question about a system; designing new interventions for, or uncovering vulnerabilities in, a system; building engagement and shared understanding and ownership of an issue amongst stakeholders; or allowing marginalised perspectives to be communicated to powerful actors.
- **System definition/boundary:** A system boundary needs to be set to make mapping focussed. The system could be a particular physical system, for example, a water catchment, or a more conceptual one, such as a policy domain. You should decide on the problem area you wish to explore. The system will then be defined by this problem or questions around it, and what impacts on it. This task is often one of the most conceptually difficult and can have a huge impact on the project.
- **Choosing stakeholders:** It is important to bring in stakeholders to cover all key parts of a system. You should consider who affects or is affected by the system; who has on-the-ground knowledge and who has a strategic overview; who is often overlooked; are there provocateurs who could usefully be invited to challenge established narrative? The process can be narrowed by reducing diversity of stakeholders, but with a cost to system representation. A useful strategy for individual sessions is to keep group size small but maintain diversity.
- **Process design:** you should begin mapping in groups to produce at least the first full version of a map. This is to ensure that the benefits of collective model building are achieved. The ideal is that a mixed group with representatives of all stakeholder communities is present for this workshop. When this is not possible, sequential workshops can be run which build on maps step by step. Many of the other steps detailed below can be performed outside workshops if it is helpful.
- **Choosing focal factor(s):** Focal factors are usually outcomes or 'functions' of the system and are the first nodes laid down in the construction of the map. Choosing the right focal factor(s) is key. They strongly affect the focus of a systems map. Try to ensure all

aspects of a system you are interested in are covered. Think about what is important to who. If some groups are not represented, you should ask participants to consider things cared about by these absent groups.

- **General factors:** once we have focal factors, we ask participants to brainstorm factors which are influenced by or influence them, and then bring these together and consolidate. The key criterion for including factors is that they make a difference to how the system works. It is important that a wide brainstorming happens so that we ensure that all domains of influence are covered.

- **Building the map:** the mapping process essentially consists of drawing causal connections between factors. Starting from the focal factors and then bringing in the general factors. The process is often staged to facilitate better system thinking. For example, in a map containing outcomes, general system factors, and policy interventions, all outcomes and the general factors which impact or are impacted by then are mapped first, with policies only added at the end. This is to prevent using familiar, but perhaps inaccurate, linear models of change.

- **Factor and connection information:** once a map structure is created, it is vital to make sure all the appropriate information on factors and connections has been collected. Some information, such as factor types (e.g. interventions, outcomes, functions), is collected throughout, but a time to gather additional information at the end is useful too. For example, which factors are controllable, by who and to what degree, which are vulnerable to particular changes, and which are 'owned' by different stakeholders. You might also ask stakeholders for any other categories they think are relevant. Equally, for connections, we should consider is there any extra information we would like to collect.

- **Verification and review:** we expect to go through several rounds of mapping followed by verification and review. Verification and analysis bring up new aspects of the system and prompt reconsideration of the structure of the map. A stable version of the map should be reached after a few iterations; however, in the longer term, maps should be treated as updatable living documents.

- **Analysis design:** analysis should be considered from the start of the process as the information collected and how maps are built is usually modified to allow the analysis that is most relevant to stakeholders. We can ask a variety of questions with analysis, for example, what are the potential influences on outcomes, are they vulnerable or supported; what are potential unexpected impacts of external change, or of planned interventions; are there co-benefits or negative indirect effects; are there trade-offs between different stakeholders' interests; are there interactions between different interventions, synergies, or clashes; are there interactions between planned interventions and other potential changes or risks. If mapping is being used to design interventions, we can use the map to consider which points of intervention would have most beneficial impacts and for who and are controllable by those wishing to act. If we want to uncover vulnerabilities, we can consider which factors have the most impact on functions that matter to different groups. When preliminary analysis is done, at every stage, stakeholders should also be asked to reflect on what is surprising or interesting in the map, and what questions to explore in further analysis.

These stages in each project will be different, and we often find we iterate between them rather than following them in a dogged sequential order.

COMMON ISSUES AND 'TRICKS OF THE TRADE'

We have bumped into many issues. Some of these we have learnt to circumvent purposefully, others we likely do subconsciously. The following issues are ones we have seen most often or felt most acutely. First, the most fundamental and most common confusion people have about the 'rules of the game' for building a PSM map is what 'positive' and (especially) 'negative' connections are. Sometimes people think of connections in a normative way, so a positive connection thus means 'this factor is good for that factor', or 'this factor influences that in a desirable direction', and conversely, a negative connection comes to mean something like, 'this factor pushes that factor in the wrong way'. It is important to be aware that this might happen and be alert to how connections are being understood and added, questioning when necessary.

A second issue during the early stages of a workshop can be the feeling that the beginning of the mapping process is slow and can start to feel like it will never be finished. This is totally normal. Early discussions on definitions and purpose, and on consolidating individual brainstorming often take up most time in a first mapping session. Once this is done, a session can rapidly accelerate so that people are building the map at a frantic pace, and the problem can flip; you end up being concerned there is not enough discussion. More generally, a sense of being overwhelmed by a map as its starts to get larger, is normal. We have heard of many people starting to use similar system mapping methods and giving up when they feel the sheer weight of the map, and uncertainty on what to do next, become too problematic. One of the most common complements we get is that we persevered when things looked stuck or too 'hairball-ish'.

On a more conceptual level, we have found it is often difficult to convince stakeholders and users, of both the value a PSM process could deliver for them, and how the analysis will work. The fact that the direction the analysis will take is not clear at the start compounds this issue. It can be hard to show examples because the analysis is bespoke, and much of its value is in iterating through version of submaps and interpretation, rather than just looking at a final version.

To overcome these, and other issues, we have regularly used the following 'tricks of the trade':

- **Be clear about the 'rules of the game' from the start**: it is well-worth spending an extra ten minutes at the start of workshop explaining the definitions of nodes and edges clearly, rather than discovering two hours in that people were misusing them.
- **Always ask people to explain**: your most powerful prompt during a workshop is 'could you explain that a bit more please?' Do not worry about looking stupid or not knowing the system in question; you are not the expert, the stakeholders are. This will encourage discussion and precision in thought and will surface assumptions and disagreements. It will help you work out when one connection would be better shown with three, or where a connection should be marked as complex.
- **Encourage people to take control**: as a facilitator, it is hard to overcome the feeling that you should be in charge of the process of building a map. Things can start this way, but it is usually possible for participants in a workshop to start to take over both the act of drawing and moving nodes and edges, and to lead the conversation (though, we have found working online this is harder to do). This should be encouraged from early in the process. Your role as facilitator thus becomes to ensure the map is sticking broadly to the rules of PSM and is coherent with any agreed aims or purposes.

- **Keep the whole in mind whilst working on details:** it is important that you keep in mind the whole system map that you are aiming for. You want a map that covers all the major areas in equivalent depth, and where distal connections are made between factors that are far apart. You will need to keep in mind the shape, balance, and connectivity as the map as a whole as well as being alert to what is happening on the microscale of local connections. 'Zooming in and out' cognitively, and prompting participants to do the same, particularly when stuck, is an important part of a mapping process.
- **Do not put pressure on the process to finish soon:** this is tricky as it involves both managing the expectations of people involved and avoiding letting yourself feel the need to finish soon. PSM is inherently an iterative process. The map will change and be refined and may never feel finished. You need to make your peace with this fact; if you chase after a false sense of completion, you will likely just feel ever further from it, and become disheartened. If you are patient, allow yourself to stop, work on something else, and return with a fresh mind, you will almost always find productive ways forward.
- **Introduce analysis early:** though the analysis stage does not start until you have a map you do not think is going to change lots anytime soon, it is useful to introduce the idea of analysis early and gather ideas for what could be done. Show people the sorts of analysis that are possible. This allows people involved to have a sense of one of the key ways you will use the map and allows you to tailor the process considering their comments and ideas. The reconfigurable nature of the analysis means that people can often come up with their own ideas once they have grasped the general principles.

WHAT IS PARTICIPATORY SYSTEMS MAPPING GOOD AND BAD AT?

PSM is likely to work best when we want to use systems mapping in a relatively participatory, inclusive, and flexible manner, but when we also want the structure given by clear definitions of how the model works and how it can be analysed. Importantly, it offers this formalism in contexts where we either don't have data to set up or validate a model, or where we are not confident we have sufficient understanding to turn a system map into a dynamic simulation. It is also directly aimed at capturing the full complexity of a system, but then finding ways to make this understanding practical and actionable using analysis. It is aimed at finding solutions to the common critique of system maps as 'horrendograms' but doing this without compromising on depth. Thus, we see PSM as sitting quite squarely in the middle of the spectrum between flexible and qualitative methods, and more formal quantitative methods.

Where PSM is less useful is when we find ourselves wanting to be at either end of this spectrum. If we want a method that is highly intuitive and allows people to engage on any level they wish, PSM will be inappropriate. Conversely, if we want to quantify our model, or have a dynamic model, PSM won't be the best bet. Probably its single biggest weakness is its inability to explore dynamics in systems; it is a relatively static method.

A Brief History of Participatory Systems Mapping

Our work with systems mapping really started with Fuzzy Cognitive Mapping (FCM) and the project presented in Penn et al. (2013). We found the overall approach of FCM created a lot of value for stakeholders and opened interesting avenues for research. However, we had concerns about how the analysis in FCM is sensitive to many subtle assumptions. In particular, the functions specified for how a change in one factor affects the next. Being sensitive to assumptions is not in itself a problem, this is true of many modelling approaches, but here, assumptions made by the researcher often had a greater impact on results than stakeholder input. Great care must thus be taken for analysis to produce outputs that were meaningful, and the risk of overinterpreting model artefacts increases. We felt this had the potential to constrain the participatory element of FCM and limit the co-design of the analysis. Thus, we started to explore the question of how we could analyse these types of complex system maps in a way which was less sensitive to assumptions, but which was also more transparent, intuitive, and participatory in nature.

The idea of using network analysis and causal flow emerged (see Penn et al., 2016) and this has been refined and extended through numerous projects since. These projects have tended to be with stakeholders from the public sector, so the approach has naturally pivoted to thinking about interventions and their outcomes, despite our aim of using the method to instil wider systemic thinking. This has become one of the main sites of innovation in map building, finding ways to encourage system-wide thinking and analysis, while keeping time-poor and objective-focused stakeholders interested.

Getting Started with Participatory Systems Mapping Yourself

So far, we have avoided the topic of software to use for PSM, however, picking the right software for you will be key to your success. There is a wide array of options, which fall into three types: general purpose diagramming software; network visualisation and analysis software; and purpose-built software. Table 5.2 outlines some of the pros and cons of each

Table 5.2 Software options for PSM. Source: Authors' creation.

Option	Examples	Pros	Cons
General purpose diagramming software	• digarams.net • Visio • Concept board • Miro	• Very easy to use • This option is most 'human-readable' • Maps are easily editable • Maps are easily shareable • Layout easy to manipulate manually—that is, to reproduce layout from workshops	• No automated map analysis is possible, analysis can be conducted manually but is time consuming and prone to human error. • Exporting map data is possible but is often difficult (i.e. least 'machine-readable') • Cost of commercial options, but there are many free and open-source options
Purpose-built software	• CECAN's PRSM • Kumu • yED	• Easy to use • Simple and/or appealing browser interfaces • Map easily editable • Some analysis available • Maps are shareable • Some advanced functionalities	• Stability of more bespoke and less well-used software can cause issues • Analysis options are limited, but can be added on request • Cost of commercial options, but there are many free and open-source options
Network visualisation and analysis software	• Gephi • R and Python packages	• Full range of analysis • Can automate analysis approach once developed	• Steep learning curve to use • Shareable as a file, but not possible to work on at same time as others • Layouts generated rely on algorithms, and manual manipulation is difficult or impossible. Layout will be unfamiliar to stakeholders

of these and mentions some examples. In practice, you might want to use two or more pieces of software for different purposes; if you do, you should have a plan for how you will export your map from one to the other—you do not really want to have to manually create your map twice.

Here are a few further resources we would recommend:

- **Detailed workshop guide:** Penn and Barbrook-Johnson (2019) outlines a detailed guide to facilitating a workshop, covering all the key stages you may need to include. If you are planning a workshop, we recommend using and adapting it to your needs.
- **Detailed Process Design Guide:** Penn and Barbrook-Johnson (2022) provides in-depth guidance on how to design a full PSM process to best fit with a system and stakeholders' challenges and blind spots.
- **Centre for the Evaluation of Complexity Across the Nexus (CECAN) PSM briefing notes:** there are three short and accessible write-ups of some of the larger projects we have used PSM in, in briefing notes published by CECAN (see Barbrook-Johnson, 2019, 2020; Barbrook-Johnson et al., 2021). These can be useful to share with prospective participants.
- **Academic examples:** a subset of these is written up in Barbrook-Johnson and Penn (2021) in longer academic journal paper form.

You should now have everything you need to get out and start your own PSM project. We cannot emphasis enough, your priority should be ensuring you have good engagement from a set of stakeholders or a user for the work. It is important to get this lined up before doing much else because the aims and design of the project will depend on them. Beyond that, we would recommend a key first step is finessing your thinking about how a PSM process is going to be of value to them, how it will connect to their existing work, and how you are going to design it to maximise chances of success. Think about these sorts of questions before you think about the methodological details of map construction and analysis, so that the real strength of this method—bespoke map construction and analysis design that provides meaningful and actionable insights—can be fully exploited. Good luck!

REFERENCES

Barbrook-Johnson, P. (2019). *Negotiating complexity in evaluation planning: A participatory systems map of the energy trilemma.* CECAN EPPN No. 12. https://www.cecan.ac.uk/resources

Barbrook-Johnson, P. (2020). *Participatory systems mapping in action: Supporting the evaluation of the renewable heat incentive.* CECAN EPPN No. 17. https://www.cecan.ac.uk/resources

Barbrook-Johnson, P., & Penn, A. S. (2021). Participatory systems mapping for complex energy policy evaluation. *Evaluation, 27*(1), 57–79.

Barbrook-Johnson, P., Shaw, B., & Penn, AS. (2021). *Mapping complex policy landscapes: The example of 'Mobility as a Service'.* CECAN EPPN No. 18. https://www.cecan.ac.uk/resources

Bromwich, B., Penn, A. S., Barbrook-Johnson, P., & Knightbridge, J. (2020). Systems analysis for water resources: Final report. Defra report. http://randd.defra.gov.uk/ (search for WT1512)

Penn, A. S., & Barbrook-Johnson, P. (2019). Participatory systems mapping: A practical guide. CECAN toolkit. https://www.cecan.ac.uk/resources/toolkits/

Penn, A. S., & Barbrook-Johnson, P. (2022). How to design a Participatory Systems Mapping process. CECAN toolkit. https://www.cecan.ac.uk/resources/toolkits/

Penn, A. S., Knight, C. J. K., Chalkias, G., Velenturf, A. P. M., & Lloyd, D. J. B. (2016). Extending participatory fuzzy cognitive mapping with a control nodes methodology: A case study of the development of a bio-based economy in the Humber region, UK In *Including stakeholders in environmental modeling: Considerations, methods and applications,* Ed. S Gray, S Gray, R Jordan, M Pallisimio

Penn, A. S., Knight, C. J. K., Lloyd, D. J. B., Avitabile, D., Kok, K., Schiller, F., Woodward, A., Druckman, A., & Basson, L. (2013). Participatory development and analysis of a fuzzy cognitive map of the establishment of a bio-based economy in the humber region. *PLoS ONE, 8*(11), e78319. https://doi.org/10.1371/journal.pone.0078319

CHAPTER 6

Fuzzy Cognitive Mapping

Abstract This chapter introduces Fuzzy Cognitive Mapping (FCM), a method for developing and analysing 'semi-quantitative' (i.e. using and producing indicative rather than predictive numerical values) causal models. We explain in simple language what an FCM map is made up of and the two main ways in which analysis is done. We go into some detail on how to do it yourself and provide reflections on common issues and tricks of the trade. We also discuss its roots and debates in the field since. Finally, we provide some advice and resources for getting started yourself.

Keywords Fuzzy Cognitive Mapping • Cognitive mapping • Mental models

Fuzzy Cognitive Mapping (FCM) holds a special place in our hearts, being one of the first system mapping methods we used together. We found the intuitive nature of FCM map building, and the offer of 'quick and dirty' exploration of a system's dynamics, appealing. Over time, our practice, particularly in how we analyse system maps, has shifted in response to some of the drawbacks we felt the method had. It is hard for us to write this chapter without that history in mind. We worried about this causing bias, but our research for this chapter has led us to understand that the same debates we had about what was appropriate to do in analysing FCM maps have been at the centre of FCM debates more widely. To reflect this, we present both the schools of thought in how to analyse FCMs.

© The Author(s) 2022
P. Barbrook-Johnson, A. S. Penn, *Systems Mapping*,
https://doi.org/10.1007/978-3-031-01919-7_6

In the rest of this chapter, we describe what FCM is in a comprehensive but as simple as we could manage way. We describe any variety in terminology or practice we have seen, and we describe in detail how to do FCM. After the 'how to' section we discuss common issues and tricks of the trade. We then reflect on what the method is good and bad at, before outlining the history of FCM. Finally, we outline some practical next steps for readers wanting to use it themselves.

WHAT IS FUZZY COGNITIVE MAPPING?

FCM involves building a model of a system made up of boxes and connections. The boxes, normally referred to as 'factors' or 'concepts', can represent anything which is expressed as a variable (i.e. it makes sense to think of it going up or down). Importantly, it does not have to be quantifiable or to have data behind it which means it can be used to capture knowledge without empirical data. The connections (i.e. arrows) are often referred to as 'edges', and they represent causal links between factors. They do not represent a vague notion of 'something is going on here' but are meant to show direct causal influence. The exploration of how these causal influences propagate through a system when it is subject to change or intervention is at the heart of what an FCM is intended to do. There are, however, two main approaches to this in common use within the community, and it is often not explicitly stated which approach is being used.

The first approach is most related to the origins of FCM and retains the same mathematical formulation. In this approach, which we will call 'causal' (following Helfgott et al., 2015), the strength of links between factors represents how certain or not we are that one factor causes, or suppresses, another. Values of factors produced in analysis represent how strongly caused (or 'activated') they are by changes in other factors based on our level of uncertainty about whether causal links in a map exist or not. Essentially, how certain we are that changes in some factors would cause changes in others, not how large those changes might be. The output of analysing a causal FCM is a list of numerical values of factors under different scenarios with each number representing how relatively strongly caused we believe a factor to be. It aims to answer the question: if we change something in the system, what implications does it have for whether the other factors in the system change?

The second approach, which we will refer to as 'dynamical' (again, after Helfgott et al., 2015), considers the propagation of effects of one factor on another, producing a simple dynamical representation of the relative magnitude of changes in factor values, that is, a plot of all factor values at each iteration. Here, magnitude of factor value does tell us how relatively large an effect is. It is used to help understand which factors and connections are most important or influenced in a system, with larger factor values interpreted as being more important or influenced/influential, and how changing map structure changes that. It can also be used to gauge how relative changes in factor values might play out dynamically, for example, whether change accelerates, stabilises, or dies away.

The differences between these two approaches to analysis are important but tricky to understand. We spend a lot of time in this section unpicking them and return to them in the history of FCM section. Table 6.1 summarises the key differences.

In both the 'dynamical' and 'causal' approaches, the factors and edges are assigned numerical values. These values and their meanings are different depending on which analysis approach is being used. If the dynamical approach is being used, factor values can take any real value. Depending on the calculations used, the values of factors can represent the size of a change in the factor values or the actual magnitude of those concepts (their initial value plus the change). Commonly, change is explored, and most factor values are initially set as zero, representing a baseline starting position (more on this in a moment). For edges in a dynamical approach, the value will usually be between -1 and +1, often in categories corresponding to 'weak', 'medium', or 'strong', which normally represents the strength of the effect of the causal influence (i.e. a value of +0.5 would mean if the source of the arrow has a value of 1, the target of the arrow will go up by 0.5 units, if the value were -0.3, with the source factor at 1, the target would go down by 0.3, and so on). (Although edge values could be given any real value without causing problems in calculation.)

In the causal approach, edge values are constrained to be between -1 and 1, while factor values are constrained to be between 0 and 1, or sometimes -1 and 1. A factor with a value of 1 is fully activated or caused, a factor with value 0 is not activated or caused. Values in between represent how certain we are that a factor is being caused by its inputs. Formally, a value of 0.5 is the most ambiguous, as it means that we have no information about whether a factor is being caused or not. The value of edges represents our certainty about whether factor A causes factor B. If the magnitude of the value on the link is 1, we are certain that A causes B; however, the lower the magnitude, the less certain we are that A actually causes B. The sign on the link represents positive or negative causation: a

Table 6.1   Comparing the 'causal' and 'dynamical' approaches to analysing FCMs

Issue	Causal approach	Dynamical approach
Interpretation of arrows	Represent our certainty about whether factor A causes or suppresses change in factor B	Represent the magnitude of influence of factor A on factor B
Initial values of factors/concepts	Set to either 0 or 1 depending on role in analysis/scenario	Can take any real value, but often set to 0, or to 1 if a driver or investigating impact of change in that factor
Value of edges	Must take a value between -1 and 1	Can take any real value, but typically kept between -1 and 1
Intuition behind the purpose of analysis	Assessing the strength of certainty that factors are caused or suppressed by changes in the system. Considers, if we change something in the system, what implications does it have for the causation of other factors in the system	Assessing the relative changes in the magnitude of factor values in the system under different change scenarios. Considers how much (relatively) will magnitudes of different factors be affected by change and thus how much they are influenced or influence others
Key constraints in analysis	Thresholding or squashing function must be applied to factor values to keep them between 0 and 1 or -1 and 1	Thresholding function not required
Outputs of analysis	Ranking of factors in terms of how much they are actively caused or suppressed by a change in the system	Plot showing the relative values of factors through 'iterations' of the analysis. Or ranking of factors in terms of how much they influence or are influenced
Pros	Consistent with original approach	Intuitive and appealing output. More intuitive map building
Cons	Counter-intuitive interpretation of arrows and factor values. Does not tell us anything about how much something might change. Results sensitive to form of squashing function chosen	In practice, has created confusion about the appropriateness of interpreting the edges and values in this way. Dynamical output can be misinterpreted as a simulation
Background	Reflects original FCM maths and interpretation in Kosko (1986)	Reflects wide adoption of FCM in participatory mode with a more intuitive interpretation of maps

Source: Authors' creation

positive link from A to B means that we believe A causes B to happen, and a negative sign means that A causes B not to happen.

Figure 6.1 shows an example of an FCM representing deforestation in the Brazilian Amazon. This map is presented in Kok (2009) and is developed to explore possible futures of deforestation in the Amazon. For comparison, Fig. 6.2 shows an FCM of the UK Humber region bio-based economy from some of our work (Penn et al., 2013). Both maps were constructed to fit with a dynamical approach. We can see how the maps look quite different but contain much the same sort of information. The Humber map is labelled with connections as positive or negative, and then weak, medium, or strong. The Amazon map has the numerical values annotated next to the arrows. Note that drivers (i.e. factors with no incoming connections from within the map) are here given a self-reinforcing loop to prevent their value going to zero during analysis. A map developed for a causal approach would look similar, but without self-reinforcing loops.

Analysis of maps usually involves comparing a baseline scenario, that is, the map as it is under the influence of external drivers or just its own

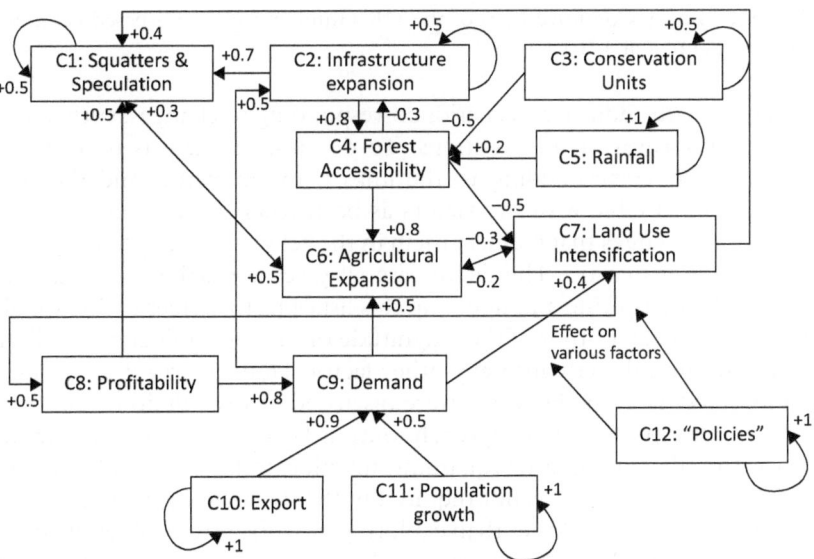

Fig. 6.1 Fuzzy Cognitive Map of deforestation in the Brazilian Amazon. (Source: Kok (2009))

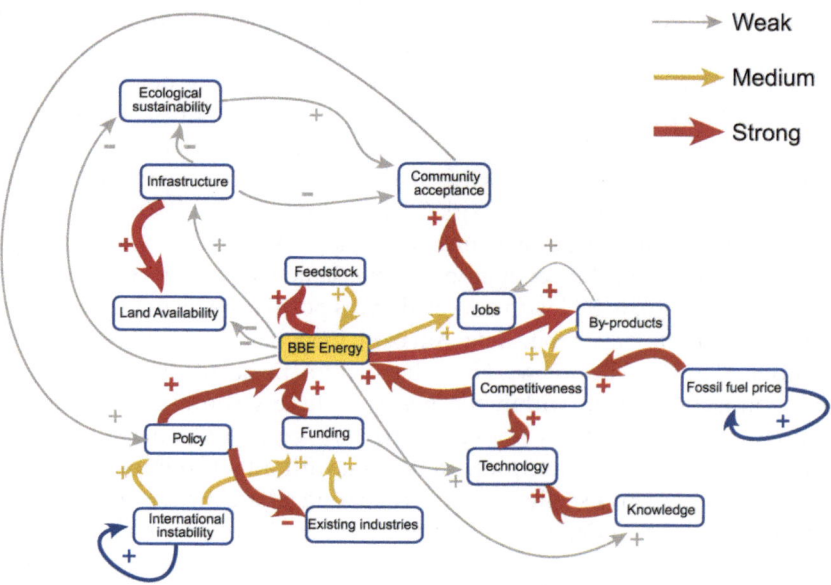

Fig. 6.2 Fuzzy Cognitive Map of the UK Humber region bio-based economy. (Source: Penn et al. (2013))

structure, with additional scenarios corresponding to changes, such as new external pressures or interventions. To do this, the map is transformed into a matrix corresponding to the links between factors and the edge weights (i.e. a table with the factors as both rows and columns, and the values on the edges that connect them in the cells where they 'cross'). This is the 'weight matrix'. The factors are all given a starting value, and by repeatedly multiplying and updating this list of factor values by the weight matrix, we get an output of the magnitude of the total influences of all the factors feeding directly into any other factor. Don't worry if this sounds complicated: it is; the FCM software discussed below will do this for you. Importantly, in the causal approach, output factor values are often modified with a thresholding or squashing function to keep values between 0 and 1 or -1 and 1. This is not required in the dynamical approach.

For the most part, rather than exploring how the actual values of factors or concepts change, FCM analysis explores how changes in a few factors propagate through the system. In the dynamical approach, a baseline scenario is often then produced by setting the initial values of any drivers in the map to 1 or less and the other values to zero. The model output is then calculated, propagating the change through the map using the values of edges and factors. This process of updating factors' values based on other factors' value changes and edge values is iterated until such point that a stable pattern can be seen in the values of factors. That is either values no longer change, or they change in a repeating cycle.

An example of the types of output this process produces can be seen in Fig. 6.3. The top left plot is the output from the map shown in Fig. 6.1; the other three plots are outputs from slightly modified versions of this map structure intended to represent different policy scenarios. This output is dynamic in the sense that it is based on propagating changes to factor values; however, it is important to remember it does not represent changes through time. Rather, FCM practitioners often refer to changes through 'iterations' of the map. As this is a 'semi-quantitative' method, interpretation involves comparing only the relative ranking of factor values.

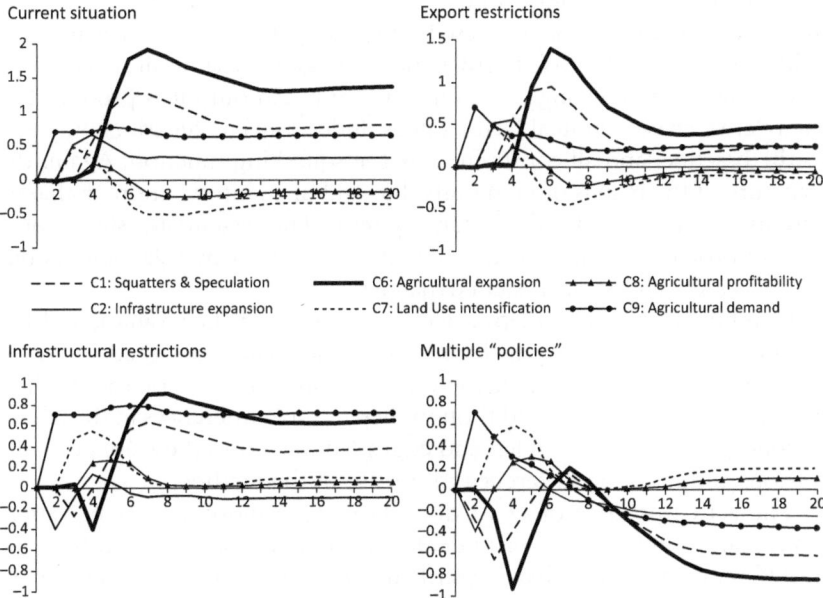

Fig. 6.3 Example outputs from an FCM dynamic analysis. (Source Kok (2009))

There tend to be four types of outputs of the dynamic analysis: (i) the factor values all change and stabilise at new values; (ii) the factor values change and continue in one direction in a 'runaway change'-type scenario; (iii) values change but return to zero; or (iv) values change up and down in cycles. Looking for these types or calibrating the model to produce them can be a useful discussion tool in workshops. Calibration to achieve this involves tweaking edge weights, in particular, modifying the strength of short feedback loops which have a strong impact on the dynamics. Sensitivity testing may be done to determine which links cause the system to destabilise, but often an experienced modeller will have a sense of what sort of structure will stabilise or not and keep this in mind during map construction. This stage is the most technically demanding, and we recommend you practise and ensure familiarity with it before building maps you plan to use in your research.

In the causal approach, analysis tends to be focused on comparing the impacts of changes in the map on relative factor values. The factors are initialised, and the model run to equilibrium to produce a baseline. To examine different scenarios, one or more of the factors in the map has its initial value increased or decreased and 'clamped' at this new value (i.e. held artificially at this point rather than changing dynamically). This represents an external change or an intervention. An output is produced of the relative change in final factor values under a given scenario, compared to the baseline, and all constrained by the squashing function. Again, the structure of the map is often modified to add in an intervention or change and its connections to the extant system. This essentially shows how change propagates through the system as perceived by stakeholders but gives no idea of possible system dynamics.

Combined, the static map of factors, edges, and their values, and the dynamic or causal analysis based on these values is what FCM is. Terminology is fairly settled, there are no other names for the method which are regularly used, and though there can be differences in the names of concepts/factors/variables, and edges/connections, these are normally obvious and make no difference to the use of the method.

The actual practice of FCM does have some more important variety which can go unreported or underappreciated. The analysis can be done in different ways, as we have described above. Moving to building maps, this can be done in one of three ways. Researchers can build them themselves on whatever evidence or input they deem appropriate; multiple maps can be built in individual interviews with stakeholders, and then combined; or one map can be built in a workshop setting with multiple stakeholders. These modes can be combined in different ways, for example, with a researcher starting a map, then running a workshop, and then conducting some follow-up interviews.

During the construction of a map there are also subtle but important differences on how a map is started and constructed, which can have profound impacts on the nature of maps built. For example, we might have a list of twenty factors, and brainstorm in a workshop the most important connections ad hoc (i.e. asking stakeholders to start where they want), drawing and building out as we connect factors; or we may systematically go through each pair of factors in a table format and consider if there is a connection. Both approaches are valid, have pros and cons, but will result in different maps. More experienced FCM facilitators will also (sometimes subconsciously) guide the process in ways which avoid potential pitfalls later (more on this below).

How Do You Do Fuzzy Cognitive Mapping?

There are three basic stages to building FCM maps. First, we develop a list of the factors to go in the map; second, we construct the map and its connections; and third, we produce the analysis and interpret it with stakeholders or users. Let's consider each of these in turn:

1. **Develop list of factors to include**: This needs to be a well-thought through list of factors which are expressed as variables. Before the list can be made, a decision and agreement must be reached on what the map will be of, that is, what is the system. This system definition can be very difficult. Though factors can be anything, they normally need to have some level of comparability in their abstraction or simplification. In practice, it is useful to limit the number of factors at this stage to twenty, simply to reduce the time required to specify connections at the next stage. In multiple individual interviews, we might provide a list we create ourselves, or let interviewees build their own map with their own factors and then synthesise the maps and factors individuals created, decide when factors are the same and pick a name, remove duplicates, or work out what to do where there are differences. This process is like coding qualitative interviews where we look for themes; it requires a lot of researcher judgement. Guiding interviews based on what has happened in past interviews

can reduce the potential for large differences and difficulty here. In workshops, this stage is normally the most time-consuming as participants must brainstorm, explain their thinking, then agree on choosing, merging, and removing duplicates from a final list. It is common to spend a large proportion of your workshop time on this stage.

2. **Construct the map**: now we can specify the connections between factors, including where arrows should go, discussion of what the process is behind them (sometimes this can uncover confusion, revealing the need for factor redefinition or the need for other arrows or factors), the direction of the arrow, whether it is positive or negative (i.e. increasing or decreasing), and what its strength is (i.e. the value of the edge weight). We do not collect the actual value or strength of the connection until the end when all connections have been drawn, so we can make sure they are all decided relative to each other, often by asking participants to rank them by strength. There can be a lot of confusion at this stage around the meaning of positive and negative connections. The method uses these terms in a mathematical sense (i.e. positive means they move together, as one factor goes up, so does the other, or as one goes down, so does the other; and negative means they move inversely, as one goes up, the other goes down, or as one goes down the other goes up), but stakeholders often use them in a normative sense (i.e. a positive influence is a 'good thing', or a negative influence is a 'bad thing'). It is also perfectly normal at this stage to relabel, add, or remove factors, as stakeholders' thinking develops.

3. **Conducting and interpreting analysis**: now that we have the static map with factor values and edge values, we can perform analysis using either of the causal or dynamical approaches described above. Because of the importance of the distinction in these two types of analysis, we have described in detail how they are done above. Here, we focus on how these are used in a participatory mode with stakeholders. Analysis in this approach, should be used as the starting point of a discussion rather than a prediction of what state the system will move to. Stakeholders can be asked whether the analysis corresponds to their understanding of the causal effects of the system as they have described it. Does this make sense to them? Do they have intuition about why is it happening? The outputs of FCM analysis in some sense provide a summary or extrapolation of the

causal thinking that the map contains. Thinking through these results considering how their own mental models of causal structure are producing them allows stakeholders to reflect on, question, and confirm or change their beliefs about this structure.

To start doing FCM you need to make a choice about the materials you use to build the map, and then the software you use to visualise and analyse it. At the building stage, the main decision is whether to use pen and paper (and probably post-it notes), or to use software straight away. Using software can seem like an efficiency saving for a map which you will need to digitise at some point; however, it comes at a big potential cost of engagement and inclusiveness when building maps in a group. Using software excludes people who are not confident using computers or unfamiliar software, and if the facilitator operates the software alone, this makes them a bottleneck on the process. In our experience, participants will likely have lower levels of engagement and discussion building a map on a computer, but there will be exceptions to this.

Once you come to digitising the map there are a variety of software options including:

- **General purpose diagramming software**: For example, diagrams. net, Visio, or yEd. There are a huge number of options here. All will be able to build the boxes-and-edges structure (hopefully, in an aesthetically pleasing way), but few, if any, will be able to run the analysis. Nonetheless, it can be useful to have a high-quality drawing of the static version of the map.
- **Spreadsheet software**: It is possible to use spreadsheet software (e.g. MS Excel, LibreOffice Calc) to implement the dynamic analysis of a map with the formula functionality. This will also quickly produce new plots for you, once you have set them up. However, you won't get a nice visualisation of the map itself.
- **FCMapper**: This is a purpose-built piece of software developed by researchers using FCM. It is available from the FCMappers website (fcmappers.net).
- **Mental Modeller**: This is a purpose-built free browser-based software for FCM and has been used for work in many academic publications. It is based on the causal approach described above.
- **R packages, FCMapper and FCM**: For those who use R, there are a couple of packages to support developing FCM maps. There is a

steep learning curve to using R, so we only recommend this to people who already use R, or who have the time to spend on learning it and are certain they want to use FCM. Given R's flexibility, power, and large user base, it is likely to offer the most functionality for building FCM maps, though we have not used it for this task ourselves.

COMMON ISSUES AND 'TRICKS OF THE TRADE'

There are innumerable issues and questions you can come up against in an FCM project, too many to cover here in detail. Instead, we will consider some of the more salient and conceptual issues that apply to FCM.

With FCM it is often the dynamic analysis that users and stakeholders are most drawn to. Although all FCM practitioners refer to their method as 'semi-quantitative' and emphasise that it is not a simulation approach, in practice, there is always a temptation to over-interpret the model output. The analysis can seem like magic (especially when done live in a workshop) and offer a false sense of certainty, truth, and scientific rigour. The outputs of any dynamic model are sensitive to the assumptions in that model, and the dynamic analysis of an FCM map built based on a participatory process is the same, but those assumptions have come from a place of (quick and dirty) consensus building and group deliberation, rather than cold methodical modelling. Additionally, the simple nature of the mathematics itself, whilst allowing rapid modelling, can hide unexpected pitfalls. In a dynamical FCM, edge weights and the presence and strength of feedback loops drive the output. The process of model calibration, altering these connections to obtain an equilibrium is intended to make the model output more interpretable by stakeholders, not to build a more accurate simulation.

If, on the other hand, a causal FCM approach is being used, with threshold functions on the factor values, then the model output is extremely sensitive to the mathematical form of these functions. In fact, when using this mathematical formulation, changing the form of the threshold function can have more of an impact on the results than the map structure, even to the degree of reversing model results. It is crucial therefore to do extensive sensitivity analysis when using a causal approach. We do not mean to suggest the model output has no value; it does. However, it is vital that it is used in the right way (i.e. as a discussion and thinking tool rather than as a forecast of what might happen), and those being shown it understand these caveats properly.

These, and other, common issues can normally be dealt with careful planning, an eye for detail, and a reflexive approach. Iterating through map building, analysis, and interpretation will also be invaluable in ironing out issues. Iteration generally is a key theme to many of the approaches in this book—you will be sick of us talking about it soon!

There are also some tricks of the trade for FCM which can unlock potential issues. The first, is to think about the analysis right from the start of map building. This is tricky at the beginning, but with practice you start to intuit whether a map structure will produce interesting outputs or not. As you build, look for feedbacks; are there any? Are there many short loops (i.e. with factor A affecting B and B affecting A)? Ideally you want to have a map with feedback loops, but these need to be questioned carefully so they reflect reality and beliefs as best as possible, not just dropped into the map without much thought. Direct loops should always be questioned, are there intermediate factors which we should route a feedback via? A short, direct feedback will likely have a strong effect on the dynamic analysis, can we specify it so there are more steps?

The ability of FCM maps to include different types of factors is appealing. However, this can lead to issues if factors operate on different timescales. For example, if factor A influences factor B over many years, and factor C also influences B, but on a daily basis, the analysis of the map will treat them in the same way, which may produce misleading outputs. One solution may be to make the long-term influence take a small value though this won't always make sense. A more reliable solution is to try to redefine the boundary of the map, and thus the system being considered, so that factors with very different timescales do not have to be artificially included together.

Much of the value of FCM is in a process of capturing and then challenging stakeholders' mental models. In the dynamical approach, often altering the map live in a workshop to expose ways in which simple changes in system structure can alter outputs. Although this takes much experience to do live, a simple trick to try is to check some of the key outcome factors in the map and to ensure these are not represented as sinks (i.e. only with incoming arrows). Producing maps with this structure is common but leads to boring dynamics and is unlikely to reflect reality.

What Is Fuzzy Cognitive Mapping Good and Bad At?

FCM's biggest strength (but also arguably the biggest risk in using it) is in its intuitive and quick descriptions of a system's structure and semi-quantitative outputs. It does not require data or empirical evidence to inform or validate the map and its analysis. It is a form of quantitative storytelling, bringing together stakeholders and their narratives in a way which allows stakeholders to question and perhaps change their own assumptions about the system. The quick and dirty approach makes it ideal for a workshop context and use with stakeholders, and its ability to capture qualitative and quantitative aspects of a system mean that anything that matters on the ground can be included, making the method inclusive. The dynamical analysis, as well as being engaging and exciting to see, extrapolates and makes visible to stakeholders the logical implications of their beliefs about system structure, as it shows what factors are more highly driven, influenced, or caused by external changes or interventions. Essentially allowing them to ask, 'does this factor or change actually cause what I think it does according to what I believe the system structure to be?'. This can lead to profound learning experiences.

The nature of the analysis required to work in this quick and dirty way, however, is the source of many problems. It creates the possibility of the analysis being easily misapplied or misunderstood. At worst, meaningless, but quantitative, analysis could be taken as truth by users and stakeholders. Although the dynamical analysis is often cited as a key hook, many practitioners say that the most important output is the map itself.

A Brief History of Fuzzy Cognitive Mapping

FCM first appeared with Bart Kosko's (1986) paper outlining a modified approach to cognitive maps (themselves first outlined by Axelrod, 1976), applying fuzzy causal functions to connections (i.e. using –1 to 1 to represent the certainty of presence of a + or – causal link). Cognitive maps had been around for a little while and had been applied in a variety of domains; they had binary values for connections (i.e. either on or off, there or not) and were among the first representations of causal connections between factors identified by stakeholders, rather than researchers alone. Kosko's contribution in adding the 'fuzzy' was two-fold; to allow a more nuanced representation of causal reality, and to allow the maps to be 'computed' (i.e. with numerical values, to allow values of factors and edges to be combined to calculate the outcome of the values and the map structure).

Initially, most FCM research was reported in technical journals, with focus on the technical and mathematical details and methodological development. This meant FCM was slow to reach new audiences and potential application areas. While there is still much highly technical FCM work, since 2000, there has been both an increase in the number of published FCM works, and a spread into domain journals and a variety of fields, from social sciences through medicine and natural sciences. There does appear to be a clear divide between (i) work which uses FCM in a participatory mode, treating maps primarily as discussion and learning tools; and (ii) work which focuses on model discovery/inference/learning from data and treats maps as a form of neural network. Jetter and Kok (2014) provide a detailed history, with reference to many examples, for readers wanting to explore this history more fully.

Over time, however, as this method has been used more in participatory contexts, an ambiguity about whether certainty of cause and activation, or strength of causal link and magnitude of change, in factors is being investigated. Many authors use a causal mathematical modelling approach, but discuss strength of connections and size of effects, or switch between causal and dynamical terminology. This is potentially problematic as we are not actually calculating how values of factors change with this formulation and so a danger exists that a non-sensical model is produced and hence misinterpreted. A consistent causal approach could be used within stakeholder workshops by mapping certainty regarding cause rather than strength of effect on links. However, the ideas of factors being caused or not caused are by no means as intuitive as thinking about factors having an actual value or link weights representing a size of effect of one factor on another. This could cause issues within workshops when soliciting values for the map and indeed seems to cause issues in the literature. These issues are discussed in detail and clearly explained in the excellent working paper by Helfgott et al. (2015).

Getting Started with Fuzzy Cognitive Mapping Yourself

There is a noticeable dearth of resources and guides for getting started with FCM. However, we would recommend the following reading:

- Jetter and Kok (2014) provide a relatively accessible history and introduction to FCM, as well as a detailed discussion of how to 'do it', including consideration of how to design a process, not just the method itself.
- Özesmi and Özesmi (2004) provide a detailed description (which sometimes errs on the technical side) of how to apply their (highly cited) approach to using FCM, including exploring policy scenarios.
- https://www.mentalmodeler.com/ as well as access to the software, the mental modeller website introduces FCM and several case studies of its use.
- Felix et al. (2019) provide a more technical review of FCM, including discussion of software options.

Beyond getting a firmer and more detailed grip on how to use the method, we would recommend two things. First, decide on the materials and software you are going to use to help you do FCM. This is an essential choice and making it will allow you to experiment with the method (either making 'test' maps in small groups with colleagues, or playing with different software) to get a real hands-on feel for it. Second, you need to find the windows of opportunity to apply FCM usefully. If you want to use it with stakeholders, speak to them and find out how it might be of value, what are the questions they have that might be amenable to FCM, what are the processes and workflows which an FCM project might feed into.

References

Axelrod, R. (Ed.). (1976). *Structure of decision: The cognitive maps of political elites.* Princeton University Press. https://doi.org/10.2307/2616237

Felix, G., Nápoles, G., Falcon, R., Froelich, W., Vanhoof, K., & Bello, R. (2019). A review on methods and software for Fuzzy Cognitive Maps. *Artificial Intelligence Review, 52*(3), 1707–1737. https://doi.org/10.1007/s10462-017-9575-1

Helfgott, A., Lord, S., Bean, N., Wildenberg, M., Gray, S., Gray, S., Vervoort, J., Kok, K., & Ingram, J. (2015). Clarifying fuzziness: Fuzzy Cognitive Maps, neural networks and system dynamics models in participatory social and environmental decision-aiding processes. *Processes EUFP7 TRANSMANGO Working Paper* 1.

Jetter, A. J., & Kok, K. (2014). Fuzzy Cognitive Maps for futures studies—A methodological assessment of concepts and methods. *Futures, 61*, 45–57. https://doi.org/10.1016/j.futures.2014.05.002

Kok, K. (2009). The potential of Fuzzy Cognitive Maps for semi-quantitative scenario development, with an example from Brazil. *Global Environmental Change, 19*(1), 122–133. https://doi.org/10.1016/j.gloenvcha.2008.08.003

Kosko, B. (1986). Fuzzy Cognitive Maps. *International Journal of Man-Machine Studies, 24*(1), 65–75. https://doi.org/10.1016/S0020-7373(86)80040-2

Özesmi, U., & Özesmi, S. L. (2004). Ecological models based on people's knowledge: A multi-step Fuzzy Cognitive Mapping approach. *Ecological Modelling, 176*(1–2), 43–64. https://doi.org/10.1016/j.ecolmodel.2003.10.027

Penn, A. S., Knight, C. J. K., Lloyd, D. J. B., Avitabile, D., Kok, K., Schiller, F., Woodward, A., Druckman, A., & Basson, L. (2013). Participatory development and analysis of a Fuzzy Cognitive Map of the establishment of a bio-based economy in the Humber region. *PLoS ONE, 8*(11), e78319. https://doi.org/10.1371/journal.pone.0078319

CHAPTER 7

Bayesian Belief Networks

Abstract This chapter overviews Bayesian Belief Networks, an increasingly popular method for developing and analysing probabilistic causal models. We go into some detail to develop an accessible and clear explanation of what Bayesian Belief Networks are and how you can use them. We consider their strengths and weaknesses, outline a brief history of the method, and provide guidance on useful resources and getting started, including an overview of available software.

Keywords Bayesian Belief Networks • Bayesian networks

Bayesian Belief Networks (BBNs) were a real no-brainer to include in this book. Though they have been around for a little while now (30 years or so), we have seen a growing interest and excitement about them in a range of different fields, many concerned with social and policy questions. The roots of BBNs are in some of the more technical academic fields, such as computer science and statistics, but recognition of the value they can deliver has spread to many domains. There is a range of software options (more on these at the end of the chapter) to help you use BBNs, which mean that you don't need to have a deep understanding of the mathematics behind them. However, importantly, these don't give you 'too much power' without understanding, to risk doing genuinely inappropriate or misleading analysis. BBNs also fill an important niche in the landscape of different methods we explore in this book. They give us a method which

© The Author(s) 2022
P. Barbrook-Johnson, A. S. Penn, *Systems Mapping*,
https://doi.org/10.1007/978-3-031-01919-7_7

has some of the best bits of quantification, allowing us to 'put numbers on things' while also incorporating uncertainty in a meaningful way.

The term 'Bayesian networks' was coined by Judea Pearl in the late 1980s. He is an interesting and vocal character (his Twitter account is well worth a follow for the illuminating debates he instigates), and his body of work on causal inference has had a growing influence in recent years, most notably in economics, quantitative social science, and social data science. If you find this chapter interesting, and are looking for something more formal, we recommend also doing some reading around directed acyclic graphs (DAGs) and their use in causal inference.

Because of BBN's roots in computer science, there is a large literature on their use based around the idea that we can use learning algorithms to generate their networks (or maps) directly from data. In common with Fuzzy Cognitive Mapping, there is a parallel literature on developing them in more participatory modes, from stakeholder knowledge. We will focus more on the latter in this chapter, though we will touch on the concepts behind developing BBNs, and other system map types, from data in Chap. 9.

As in all the methods chapters in this book, we will now give as clear and accessible a description of what BBNs are and how to do them, as we can muster. We will avoid the use of mathematical notation to do this, there are other introductions which cover the maths well (e.g. Neapolitan & Jiang, 2016). We will then explore some common issues and tricks of the trade for using them, try to pin down what they are good and bad at, before outlining a brief history of the method and pointing to some key resources for getting started yourself.

What Are Bayesian Belief Networks?

BBNs, as with all networks, are made up of nodes (which here, represent variables, factors, or outcomes in a system) and edges (which represent the causal relations between these nodes). So far, so familiar, and like many of the methods in this book. What sets BBNs apart is that each node has some defined states (e.g. on or off, high or low, present or not present) and some associated likelihoods of being in each of those states. These likelihoods are based, in probabilistic fashion, on the states of the nodes that they are connected to, that is, the nodes from which they have arrows going into them. In the language of probability, nodes are 'conditionally dependent' on the states of the nodes that they have a causal relationship with. So, the BBN is the network and the collection of conditional probabilities (usually shown in simple tables or plots annotated onto a network diagram), denoting the likelihood of nodes taking different states. The last key point to mention is that BBNs are acyclic; that is, they do not have any cycles or feedbacks, and the arrows must flow all in one direction. This is an important distinction between other methods which do have cycles, such as Causal Loop Diagrams and Participatory System Maps.

It is useful to make this more tangible quickly, so let's look at an example. Figure 7.1 shows a simple BBN of the effects of 'rainfall' and 'forest cover' in a river catchment, and the links through to some different outcomes, such as 'angling potential' and 'farmer income'. This is a simple BBN, you can see the acyclic structure, and focus on outcomes we or others might care about.

The same BBN, but this time with the states of each node shown along with the probability distribution for them is shown in Fig. 7.2. This allows us to see, for example, if 'rainfall' and 'forest cover' are both high, the likelihood of all the other nodes taking specific values.

We could explore different scenarios by setting the states of 'rainfall' and 'forest cover' differently and seeing how this affects the rest of the map. This is a common way of using BBNs, setting certain node states given our observations (or hypothetical scenarios we are interested in), and seeing what this implies about the probability of states in other nodes. With 'rainfall' and 'forest cover', we are setting values at the 'top' of the network (sometimes referred to as 'root nodes' or 'parent nodes', i.e. with

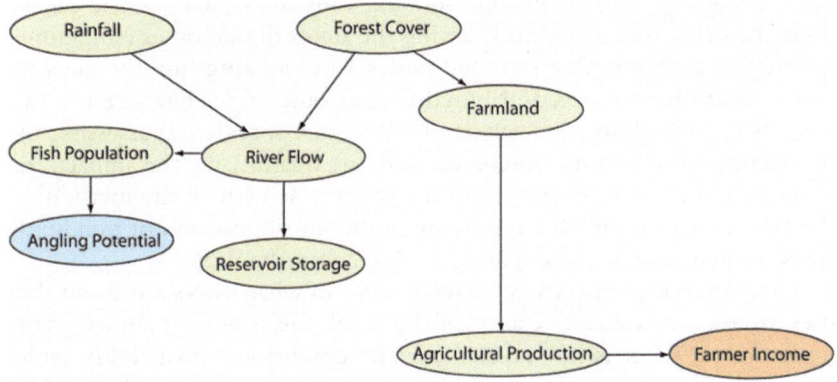

Fig. 7.1 An example simple BBN. (Source: Bromley (2005))

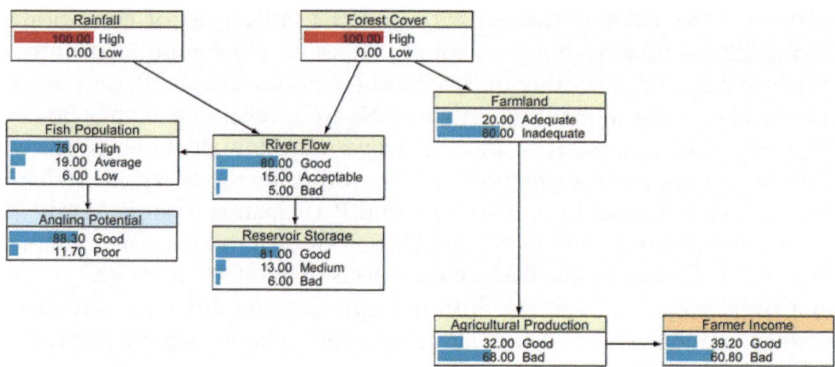

Fig. 7.2 An example BBN, now with nodes states and probability distributions. (Source: Bromley (2005))

Table 7.1 An example conditional probability table based on the reservoir storage node in the BBN in Figs. 7.1 and 7.2

		River flow		
		Good	Acceptable	Bad
Reservoir storage	Good	0.9	0.6	0
	Medium	0.1	0.3	0.1
	Bad	0	0.1	0.9

Source: Bromley (2005)

no nodes going into them) and looking causally 'down', but it can be done the other way round too; setting the states of outcomes (sometimes referred to as 'leaf nodes' or 'child nodes') and looking 'up' the network to see what might have contributed to that outcome. These are the two main types of insight the analysis of BBNs can provide: (i) assessing the probability of achieving outcomes, and (ii) quantifying the impacts on outcomes of changes elsewhere in the system. As with all the methods in the book, there is also huge potential value in the process of building a BBN, to generate discussion and learning about the topic.

These examples help us get a quick sense of what BBNs are about, but they are focused on the 'results' of the BBN; this is what is shown in the node tables. What is not present are the conditional probability tables that underpin these outputs. Table 7.1 shows what one of these tables might look like for one of the factors in this BBN, 'reservoir storage'. It shows different states of the parent node of 'reservoir storage', which is 'river flow', and the resulting probabilities of 'reservoir storage' taking each of its states. The numbers indicate the probability that 'reservoir storage' will take its values in the second column (i.e. if 'river flow' is good, then there is a 90% chance 'reservoir storage' is good, and 10% chance it is medium).

BBNs have some well-known and often-criticised constraints. First, they are acyclic; that is, there cannot be any feedback loops, of any length, in the network. This constraint is imposed for the calculations to work. In complex systems, it is rare for there to be no feedback loops; where there are feedback loops, these are often powerful drivers of dynamics in the system. It is possible to partially represent feedback loops by including multiple nodes for the same thing, but for different time points (we show an example of this below). BBNs that use this approach are often called 'dynamic' BBNs. A second constraint on BBNs developed with expert input is that most nodes cannot have more than two or three parent nodes (i.e. incoming connections) and nodes should not have more than a handful of states. This constraint is normally imposed so that the conditional probability tables, which are a key component of what is elicited from stakeholders to build the BBN, do not become unworkably large. By way of illustration, imagine a node with two states, and two parents each with two states themselves, this will require a 4×4 table. However, a node with three states, and three parents, each with three states themselves will need a 6×27 table. Imagine filling that in with stakeholders, cell by cell, with potentially important discussions at each step, and doing this for every node in the map. Combined, these two constraints mean that the underlying network in a BBN tends to end up being a relatively simplified model of reality compared to some of the other systems mapping methods in this book. This is not necessarily a problem, but it is a constraint we should be aware of.

These constraints often invoke ire from researchers and practitioners who want to represent whole systems and take a complex systems worldview (including from us in the past!). However, one of the typically misunderstood, or simply missed, nuances with BBN is that the use of conditional probabilities means that we can still capture some elements of the wider system in the analysis and discussion around constructing maps, even if they are not in the network explicitly. To demonstrate, consider Table 7.2. Here, we can see the probability of an outcome occurring given the state of two interventions. Even when we have both interventions 'on'

Table 7.2 Simple hypothetical conditional probability table for two interventions and an outcome

Intervention A	Intervention B	Outcome happens	Outcome does not happen
On	On	0.9	0.1
On	Off	0.8	0.2
Off	On	0.4	0.6
Off	Off	0.2	0.8

Source: Authors' creation

there is still a 0.1 probability the outcome does not happen, and conversely, when neither intervention is 'on' there is still a 0.2 probability that the outcome does happen. These non-zero probabilities represent 'everything else going on in the system'. They are often an important point of the elicitation process and allow us to capture influence on the outcome, even if we do not formally put them in the network.

You may be wondering why BBNs are Bayesian. They are referred to as 'Bayesian' because of the use of the underlying logic of Bayesian statistics (which provides a way to update probabilities considering new data or evidence) rather than because they were developed by Thomas Bayes himself. Bayesian statistics, simply put, is a field within statistics that revolves around the idea of probability expressing an expectation of likelihood based on prior knowledge or on a personal belief. This probability may be updated based on new information arriving about factors we believe to influence that event. In a sense this operationalises how our belief about a particular probability should change rationally as we learn more. This is in opposition to the Frequentist view of probability which revolves around the idea that probability relates to the relative frequency of an event. We do not want to get into the large and ongoing debates within and between these two schools of thought. However, it is important to recognise that BBNs take that Bayesian idea of probability and implement it though the network structure and conditional probability tables; the parent nodes of any node hold the prior information we are using to update our beliefs about that node. We can also use new information about the states of child nodes to update our beliefs about parent nodes using Bayesian inference.

There is a lot of variety in how BBNs are built, either directly from data, or through participatory processes with experts and stakeholders. However, the object that is produced and the analysis that is done tend to be consistent. There are extensions, such as dynamic BBNs (as mentioned above), and hybrid BBNs (which allow us to include continuous variables as well as the categorical variables we have described above). Where there is more variety is in the terminology and jargon associated with BBN. Ironically, given the formalism of the method, this is one of the methods with the highest number of different names, but less surprisingly, the opaquest technical language.

You may see BBNs referred to as any of the following: 'Bayesian networks', 'probability networks', 'dependency models', 'influence diagrams', 'directed graphical models', or 'causal probabilistic models'. We have also seen them referred to as 'Theory of Change maps' because of the similarities with these types of diagrams, that is, a focus on connections between inputs and outcomes, tendency to produce simple maps, and not include feedbacks of many causal influences. This plethora of terms seems to reflect the widespread use of BBNs in different domains rather than large differences in how they are used. Some of the key technical terms you may bump into might include 'prior probability distribution', or 'prior', which refers to the probability distribution that indicates our best guess about the probability that a state will take some value before new or additional evidence or data is taken into account; and 'posterior probability distribution', or 'posterior', which is the conditional probability we assign after taking into account new evidence or data. Note that the probability distributions assigned to states of root nodes are prior probabilities because they have no inputs on which their state is conditional.

How Do You Use Bayesian Belief Networks?

The main division in constructing BBNs is between approaches which generate the networks and conditional probabilities directly from data, using a range of different learning algorithms, and those which use stakeholder input. Here, we will focus on the latter, though we do consider the issue of developing system maps directly from data in Chap. 9. Even though we are focusing on the participatory mode of BBN, it is worth making clear that you will likely want to use purpose-built BBN software to implement your BBN. You can build the network and collect conditional probability information in standard ways (i.e. workshops, interviews) with general-purpose software or just pen and paper, but you will need the BBN software to run the analysis quickly and easily for you (more on software options at the end of the chapter).

Let's look at each of the main stages in developing and using BBNs:

- **Build the network**: you have two options at the start, working with stakeholders, you can build either a generic and intuitive causal network (i.e. which may have loosely defined factors, feedbacks, nodes with many parents) or a network which fits closely with the constraints of BBN (i.e. no feedbacks and not too many parents for any one node). Doing the former will make your workshop or interview process marginally easier and more intuitive for stakeholders, but will mean you need to convert the map you start with into a BBN form. This conversion process puts a lot of power and responsibility in the hands of the modeller or researcher, which you may want to avoid. However, it can be a useful iteration step to present back to stakeholders a 'BBN version' of a more generic system map they have built, for feedback. For building a network immediately in the form of a BBN, we would recommend you facilitate the process quite strongly, in a similar way to the Theory of Change process in Chap. 3, focusing on the key outcomes and inputs or interventions in the map, and then filling in the gaps with intermediary factors, but making sure you stick within the 'rules' of a BBN structure. You may also want to include external controlling factors and factors which are likely to directly influence an intervention's success. The MERIT guidelines document (Bromley, 2005) gives a useful and detailed guide to BBN construction.
- **Constrain the network**: if you built a more generic system map initially, as is the practice we have observed most often, you will then need to convert it into a BBN form. This will require some big decisions in simplifying the map and constraining it to have no feedbacks or have nodes with many parents (i.e. maximum two or three parents). If you feel the 'no feedbacks' issue is too problematic for your system, you may want to consider developing a dynamic BBN with nodes which represent factors at different time points. A simple example of this is shown in Fig. 7.3, where a feedback between 'wood extraction' and 'wood stored' has been added using wood extraction in two time periods. There is no formal reason to aim for a network of a certain size, but we have tended to see networks with no more than roughly twenty nodes. Given the fact you need to elicit conditional probabilities for each node, it is worthwhile trying to keep the network size small enough that this does not become an overly time-consuming process for your stakeholders.

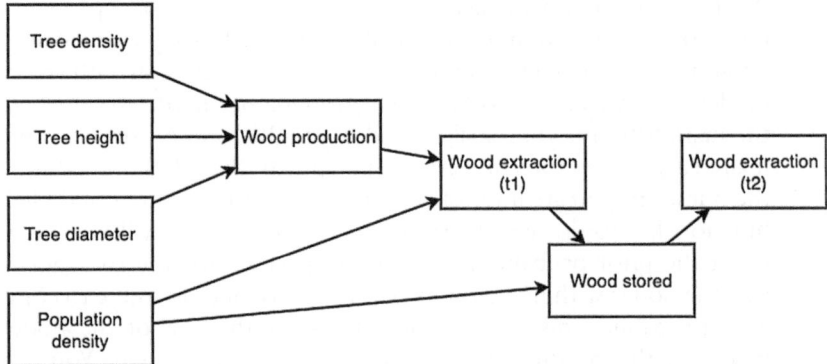

Fig. 7.3 An example dynamic BBN with a feedback between 'wood extraction' and 'wood stored'. (Source: Authors' creation based on an example in Landuyt et al. (2013))

- **Define possible node states**: once you have chosen your nodes, you need to decide what possible states all of them can take. These states need to cover not just what a variable is doing now but how it might change under any conceivable scenario in your model. They must be exhaustive, covering all possibilities, and exclusive, non-overlapping. Although variables in BBNs can be given continuous values, for the most part they are defined as discrete. The options that you have to define node states will depend on the software that you are using. These may take several forms: labels, such as low, medium, or high; Boolean variables; or numbers from a set list or numerical intervals. It is important to find a balance between a number of states that will capture or give the information you need to describe the system and the effect that has on the size of your conditional probability tables.
- **Elicit conditional probabilities**: once you have a BBN structure with all the nodes and edges defined, you need to collect conditional probabilities for every node. This is most easily done using conditional probability tables (as in Tables 7.1 and 7.2) for each node. Agreeing on these probabilities, and the number and types of states of nodes, in a workshop setting can be a key point for discussion for stakeholders. You should expect a lot of discussion at this stage, and you may need to update the structure of the map based on this.

- **Analysis**: now that you have your network and conditional probabilities, you can start doing some analysis. You will likely find value in iterating through a few rounds of analysis and discussion with stakeholders when you are working in a participatory mode. In practice, the main 'site' for your analysis will be the BBN software. It is normally easy to use the software to manipulate the BBN and address the questions you are interested in; most have simple point-and-click functionality to do this. To start with, your network will default to using the prior probabilities you have specified for the root nodes (i.e. the nodes with no incoming connections) and use these to compute probability distributions for states for the rest of the nodes based on the conditional probabilities you have defined. You can then easily specify the states (or values if you have a hybrid BBN) of any nodes in the network, and the probability distributions of the states of all other nodes will update to reflect this 'new' information. By setting states for different combinations of nodes you can create different scenarios and explore their implications in the system. The two most common types of insight we have seen this used for are: (i) to set intervention-type node(s) states and see what this implies for the chances of outcomes being achieved (i.e. what difference does turning an intervention 'on' make to the probabilities of an outcome happening?); or (ii) to set outcome-type node(s) states (perhaps based on what we have observed) and see what this implies about the states of other nodes in the network. Once you have explored the BBN and have some analysis, there are a range of ways you can export and present this. It is common to see full networks with states and probabilities defined, as in Fig. 7.2, but others often also choose to show the probabilities of different states for key outcome nodes, given different inputs, perhaps in a simple plot or table.

The stages we have described here are generic. You will need to design a bespoke process that fits your needs. This may emphasise more engagement and iteration with stakeholders or may be more streamlined with a focus on getting to analysis and producing outputs.

COMMON ISSUES AND 'TRICKS OF THE TRADE'

As with all modelling that produces quantitative outputs, there is a common tendency to over-interpret the precision and reliability of the outputs of BBNs. It is easy for clients or other users of our BBN research to take away a few standalone numbers or plots and focus on these in isolation. Moreover, because BBN deals in probabilities, and is often touted as being useful in contexts of uncertainty (which it is), when results are taken out of context like this, they can still be interpreted as having taken into account uncertainty, creating an undue sense of confidence is using them glibly. We must encourage users to acknowledge that BBNs are always dependent on stakeholder opinion (unless developed based solely on data) and that removing outputs from that context, and not making clear either the process, or the network (i.e. the model), from which they are derived almost always dooms us to see them misinterpreted. Even in cases where outputs are not misused or misunderstood, the appeal of the diagram of a BBN with conditional probabilities annotated can also lead many to view BBN and its associated analysis as a product, rather than a process. Not recognising the value in the process of using this method is to ignore at best half its value, at worst, all its value.

The other common issue relating to the relatively formal nature of BBN is that it can make a participatory process demanding for stakeholders. The process of building a BBN can quickly become a long series of small steps, which individually are dull and/or conceptually dense. This is most common when eliciting conditional probabilities, which can create rather repetitive conversations around filling in conditional probability tables. Even during the building of the underlying network, which is typically one of the more creative and exciting stages in systems mapping, the constraints BBN imposes can become onerous and inhibit discussion, leading to frustration on the part of stakeholders.

To address these, and many other issues, we see four key 'tricks of the trade' being used regularly:

- **Emphasise iteration and learning from the process**: do not be afraid to emphasise the need for, and value in, iteration of BBN; build this into stakeholders', clients', and users' expectations. One of the reasons that iteration is so useful is that it allows us to 'fold-in' the learning we are generating as we go into the process and the products of BBN.

- **Build a generic map with stakeholders and refine later**: though you can dive straight into building a network which meets the constraints of BBN, we believe that building a more generic causal map, something like a Fuzzy Cognitive Map (Chap. 6) or Participatory System Map (Chap. 5), is likely to be the best way forward. This has the advantages of making that first major stakeholder engagement easier and more flexible, gives you the ability to refine and moderate the network so that it is amenable to the types of analysis you want to do, and creates a useful point for iteration around gathering feedback on the constrained-form network. We believe these advantages outweigh the risk of doing the network refining 'behind closed doors', reducing stakeholder engagement, or introducing researcher bias.
- **Tackle criticism of constraints head-on**: in the systems and complexity communities, it is common for people to quickly comment on the constraints BBN introduce. In the most part, these criticisms are not fatal, and often are unfair, so you should develop clear and compelling arguments as to why it is still appropriate to use BBN in your context. Useful retorts include the fact you can have feedbacks if you develop a dynamic BBN, that you can have more parent nodes if you really want to, and that the simplified networks BBN often use hide the nuance and 'capture-of-context' that can be done at the conditional probability stage (as we discuss above).

WHAT ARE BAYESIAN BELIEF NETWORKS GOOD AND BAD AT?

Arguably, all the unique strengths of BBN relate to their ability to include probabilistic representations of the states of nodes in causal networks. The mere presence of numbers is not what is useful, rather it is the way the numbers are used. Despite our warnings about abuse of outputs, it is quite difficult to produce meaningless analysis with BBN because it is based on extrapolating individual conditional probabilities (which are easy to define in sensible ways) in a relatively low-risk way, rather than turning the network into a dynamic simulation, which can be fraught with danger. Essentially, unlike a simulation, a BBN retains its transparency even whilst generating numerical output. Everything that has gone into the model is clearly visible in the network structure and probabilities. There are no 'hidden' decisions taken by a facilitator or modeller on which the outputs might depend crucially. (For example, like parameter choices in Fuzzy Cognitive Mapping or System Dynamics.)

The use of probabilities also gives us the wiggle room to bring in wider contexts and causal influences, which it may appear, from the network itself, have been simplified away. Seen in this light, BBNs are one of the better methods for capturing the effect of 'everything else going on' in the system in a meaningful way. Because variables are connected by conditional probabilities, there is no need for any mechanistic model of how they change each other's values. This means that variables of many different types can be brought together in one model, allowing a BBN to cover multiple types of domains important in a problem. Taken together, these strengths hopefully make clear BBN is most useful when we want some quantification, but where it is useful to do this in a probabilistic form, and/or where we want it to be done in a relatively understandable and transparent way.

On the flip side, there are some key weaknesses of BBNs which may be problematic in some settings. They are not good at producing whole-system views of a topic or system and will tend to lead to more simplified models. This is not only in the sense of the breadth and size of map (though size is unconstrained, it is rare to see very large BBNs, i.e. 100 or more nodes) but also in individual connections because of the constraints on the number of parent nodes any one node can have. Where we want this flexibility, or the ability to capture a whole system, BBNs will be less useful. Though dynamic BBNs can capture feedbacks, this is an awkward and potentially incomplete solution (if we only capture one or two cycles of the feedback), so in contexts where there are multiple important feedbacks, BBNs may not be the best choice. Finally, it is worth noting that BBNs can also be time-consuming to update because of the additional conditional probability information that needs to be collected for new nodes, and anything they are connected to. Though they can be updated, in situations where we want a 'live-document' or an easily updateable system map, BBNs may be a poor choice.

A Brief History of Bayesian Belief Networks

The term 'Bayesian network' was first coined by Judea Pearl in his 1988 book, *Probabilistic Reasoning in Intelligent Systems*. Together with Richard Neapolitan's 1989, and very similarly titled book, *Probabilistic Reasoning in Expert Systems*, the field as it is recognised today was born. These two books reflected much of the thinking that had been developing in the

1980s through a series of workshops (now a conference) on uncertainty in artificial intelligence. Scholars from cognitive science, computer science, decision analysis, medicine, maths and statistics, and philosophy contributed to these developments. As the name of the workshop series suggests, the underlying aim was to incorporate uncertainty into the various types of computational analyses often referred to as 'knowledge-based systems' or 'expert systems'. Put simply, to include uncertainty in these techniques, which were being designed to solve complex problems, or make complex decisions, in human-like ways. These Bayesian networks built on the earlier work of Philip Dawd in the 1970s on the representations of and reasoning with probabilistic dependences. The word 'belief' was used in the name of the method right from the start (though not consistently) because this word is used to describe assertions about probabilities in Bayesian statistics. Note, 'belief' is not used because BBNs are used in subjective stakeholder participatory processes, though this is a sometimes useful coincidence.

From the start, there were two ways to build BBN, either from data or from stakeholder input. In the early days, data was only used to define conditional probabilities (and then often checked by experts), but not the network structure. As the learning algorithms became more sophisticated, computing power more affordable, and data more plentiful, data could be used to develop the network structure too. More recently, as the value of BBN has been realised in a wider set of fields, and applied to practical policy problems, the use of BBN in a participatory mode has increased in popularity.

Getting Started with Bayesian Belief Networks

We hope you now have a clear picture of what BBN is and how you might use it. The big missing piece we have left undiscussed thus far is software. You will almost certainly want to use some BBN-specific software and there are a few options you have. Let's now outline some of these software options, and their pros and cons, in Table 7.3. Note, there are more options available than those here; these are just the ones we have used and were recommended to us:

Beyond software, there are resources we would recommend to help you to get an even more detailed understanding of BBN and how to use them, including:

- **Bromley (2005) guidance**: this free and detailed guidance goes into a lot of details on the process of using BBN, including constructing

Table 7.3   BBN software overview

Software	Pros	Cons
AgenaRisk	Commercial package, well supported. Free 'lite' version with limited functionality	Cost
BayesiaLab	Commercial package, well supported	Cost
Bayes Server	Commercial package, well supported	Cost
DPL	Commercial package, well supported	Cost Not specifically for BBN but can be used this way
HUGIN Expert	Commercial package, well supported	Cost
GeNIe	Commercial package, well supported Free research and teaching version	Cost (for businesses)
OpenMarkov	Free and open source. Comprehensive functionality	Less stable product, limited support and documentation
Netica—norsys	Commercial package, well supported. Free demo version.	Cost
UnBBayes	Free and open source, simple to use, includes a library of example models	Less functionality than OpenMarkov, limited support

Source: Authors' creation

them. The example used is water resource management, but it is useful for researchers working in any domain.

- **'Risk assessment and decision analysis with Bayesian Networks' (Fenton & Neil, 2018) textbook**: this second edition (first edition from 2013) is practical and focuses on real-world applications of BBN rather than theory.
- **'Counterfactual and causal inference' (Morgan & Winship, 2014) textbook**: this book is more focused on the methodology of causal inference, built on the pillars of counterfactuals and causal graphs (which includes BBN). Useful if you want to consider the wider picture, and theory, that BBNs sit within.
- **The Bayesian Network Story (Neapolitan & Jiang, 2016)**: this Oxford handbook article provides a detailed overview of the method, which includes a lot of the underlying maths and probability, should you want a clear introduction to that.
- **The 'classics'**: for those of you who like to return to early texts, it is well worth having a look at the two foundation books for BBN, Pearl (1988) and Neapolitan (1989).

Hopefully, you now have everything you need to get started with BBN. You don't need to have a detailed grasp of Bayesian statistics, but we would recommend you try to get comfortable with the broad ideas. Beyond this, our other priorities before starting would be to play around with one of the software packages above, and to find a couple examples of BBN in similar domains, or used in similar ways, to what you are hoping to do. Good luck!

REFERENCES

Bromley, J. (2005). Guidelines for the use of Bayesian networks as a participatory tool for Water Resource Management. MERIT. http://nora.nerc.ac.uk/id/eprint/3300/

Fenton, N., & Neil, M. (2018). *Risk assessment and decision analysis with Bayesian networks*. CRC Press.

Landuyt, D., Broekx, S., D'hondt, R., Engelen, G., Aertsens, J., & Goethals, P. L. M. (2013). A review of Bayesian Belief Networks in ecosystem service modelling. *Environmental Modelling & Software, 46*, 1–11. https://doi.org/10.1016/j.envsoft.2013.03.011

Morgan, S., & Winship, C. (2014). *Counterfactuals and causal inference: Methods and principles for social research*. Cambridge University Press.

Neapolitan, R. (1989). *Probabilistic reasoning in expert systems*. Wiley.

Neapolitan, R. & Jiang, X. (2016). *The Bayesian Network Story. The Oxford Handbook of Probability and Philosophy*. https://doi.org/10.1093/oxfordhb/9780199607617.013.31.

Pearl, J. (1988). *Probabilistic reasoning in intelligent systems*. Morgan Kaufmann.

CHAPTER 8

System Dynamics

Abstract This chapter introduces System Dynamics, a well-established method to developing full simulations of dynamic systems. We introduce the method and how it can be used to move from conceptual models to 'stocks and flows' models and simulations. We describe the steps involved in doing System Dynamics and consider common issues and tricks of the trade. We reflect on what the method is good and bad at and present a brief history of use and debates in the field. Finally, we give some advice on getting started yourself and point to some useful resources.

Keywords System Dynamics • Dynamic model • Stock and flow model

System Dynamics is probably one of the most widely known methods in this book, owing to its use in the Club of Rome commissioned report, Limits to Growth (1972). Though the report's modelling and conclusions were heavily debated and criticised, there is no doubt that it explored the dynamics of economic and population growth within the constraints of the natural world in a compelling and engaging way—it showcased the power of System Dynamics, and modelling more generally, both as analytic and thinking tools.

This high-profile example reflects something more fundamental about the method; on a surface level, System Dynamics models are intuitive and often convincing. As a simulation method which produces results over

© The Author(s) 2022
P. Barbrook-Johnson, A. S. Penn, *Systems Mapping*,
https://doi.org/10.1007/978-3-031-01919-7_8

time (results which can be interpreted as forecasts, or even predictions—more on this later), it can create a black-box, magic-like quality; the lure of the simulated world convincing people of its truth. On a deeper level, however, discerning model observers and users can see the sensitivity of model results to an array of assumptions, and how the model can be used and abused by those wishing to use it to make their own arguments. Ultimately, though much energy was used on repetitive debates around Limits to Growth, and the model itself was criticised heavily, it likely did serve its purpose well—to raise awareness of an issue, and to facilitate and support more detailed thinking and strategic insight on it. Perhaps at a cost of System Dynamics itself, it also raised much discussion and thought on the different ways to model complex systems.

The rest of this chapter follows a now familiar structure. We first describe as clearly as we can, what System Dynamics is, before outlining how it is done. In these sections, we try to capture the nuances in differences in practice and how these affect what is produced. We also cover tricks of the trade and common issues you might come up against. Next, we try to summarise what System Dynamics is good at, and what it is not so good at, and consider how people tend to react to it. Finally, having introduced the method, we outline a brief history, so readers can get a strong sense of the points of debate and the trajectory of the method's use.

What Is System Dynamics?

The short but jargon-filled definition of System Dynamics (many variants of which you will find in the literature) is 'a computer simulation method for analysis of complex dynamic systems'. If we unpick and decode this a little, we can say it (i) is typically implemented on a computer using software to carry out the calculations we put into the model, (ii) allows us to prod and poke a model to run different scenarios and look at different outputs for comparison, and (iii) is normally used to look at systems which are dynamic or moreover have some interesting dynamic properties.

More concretely, a System Dynamics model is made up of stocks and flows, and the factors which affect flows. Stocks are any entity which accumulates or depletes over time. A flow is the rate of change in a stock (usually represented by a differential equation). Stocks take numerical values, and flows are defined by equations which can be affected by any number of other factors in the model. These equations may be relatively simple, using standard operations like addition, subtraction, multiplication, division, and sometimes exponents, or more complex involving functions and parameters which moderate how variables interact. Once a set of stocks, flows, and the factors which affect them are specified in equations, the simulation can be run by choosing a starting point (i.e. a set of stock values) and computing how stock values change through repeated time steps (i.e. through being repeatedly changed by their associated flow). Figure 8.1 shows an example of a stock and flow System Dynamics model which represents tourism and pollution in the Maldives (it is taken from Kapmeier & Gonçalves, 2018). As with most of the methods in this book, the diagram is in essence made up of boxes and arrows. Common to many System Dynamics diagrams, in Fig. 8.1 we can see the stocks as rectangles, the flows as straight hollow arrows with hourglass-type symbols (these are meant to represent valves) on them, and the factors that affect flows represented by floating text and the solid (and mostly curved) arrows. The small circular feedback arrows with italic annotations are labels for the feedback loops represented by the arrows around them. B stands for a balancing feedback loop, meaning that it will tend to self-stabilise, R for a reinforcing feedback loop, which will tend to carry on increasing or decreasing. Not so common, but used here, are the dotted lines to

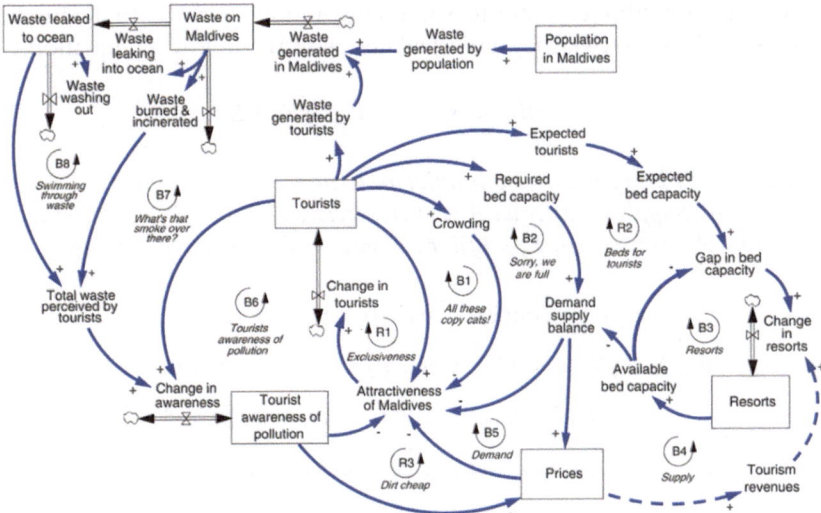

Fig. 8.1 A System Dynamics stock and flow diagram of pollution and tourism in the Maldives. (Source: Kapmeier and Gonçalves (2018))

represent influences which were not included in calibration processes due to missing data (more on calibration later).

To describe some of the equations in the model, let's take the example of the number of tourists. This is represented by the stock 'Tourists', which changes based on the flow 'Change in tourists', which is, in turn, determined by a constant historical tourist growth rate (we cannot see this in the diagram, but it is specified in text which describes the model) modified by the factor 'Attractiveness of the Maldives' (which we can see here). Essentially, the flow is a differential equation specifying the rate of change of tourists over time, as the historical rate of change is impacted by the changing attractiveness of the destination.

$$\frac{dT}{dt} = T * g_h * (1 + A)$$

where T is number of Tourists, g_h is Historical Tourist Growth Rate and A is Attractiveness of the Maldives.

'Attractiveness of the Maldives' is calculated based on five inputs: tourist numbers, crowding, demand/supply balance of beds, prices, and tourist awareness of pollution. The ways these are specified in an equation can vary from something simple to something more nuanced. In Kapmeier and Gonçalves (2018), they are calculated with the following equations:

$$Attractiveness = E_R + E_{DS} + E_P + E_p$$

where E_R = Effect of Resort Exclusivity on Resort Attractiveness; E_{DS} = Effect of Demand/Supply Balance on Resort Attractiveness; E_P = Effect of Pollution on Resort Attractiveness; E_p = Effect of Price on Resort Attractiveness

$$E_R = f_1 (Tourists / Tourists\ reference\ value)$$

$$E_{DS} = f_2 (Required\ Bed\ Capacity / Annual\ Bed\ Capacity)$$

$$E_P = f_3 (Pollution / Pollution\ reference\ value)$$

$$E_p = f_4 (Price / Price\ reference\ value)$$

where the f functions are decreasing s-shaped logistic functions, and reference values estimated from past data are used to normalise the input variables (i.e. dividing the current number of tourists by a reference value normalises it relative to that reference). The logistic functions bound the effects on attractiveness of any variable between specific upper and lower limits, ensure that increasing values of numbers of tourists, demand/supply balance, pollution, and price decrease resort attractiveness and allow the system to be set up so that a high value of any one of these factors dominates over any positive effects from others. Similar sets of equations are described for each of the stocks and flows in the map and the influences on them. These equations together allow the dynamical behaviour of the system to be simulated.

It is often assumed that System Dynamics models are deterministic using the same exact values each time and thus producing the same result every time; however, this is not true. Many models, and indeed very early examples, included stochastic elements. What types of things you can represent in stocks and flows is in theory broad. We do not have to have data on things in the model, and they don't even have to be quantifiable. Indeed, the founding figures of the method were clear that we did not need to revert to using highly quantitative factors with data to back them up. As long as it makes some sense for stocks to be talked about going 'up and down' and we can put some numbers to them and factors in equations without it being completely meaningless or inappropriate, we can make a model work.

However, in practice, people often favour more quantifiable elements and avoid 'soft' (unquantified) factors. This is perfectly understandable; the existing literature and data make it much easier to quickly implement a biophysical factor such as 'rainfall', for example, compared to a social factor such as 'Attractiveness of the Maldives' (as above). The natural and physical sciences tend to have more data and quantitative models of almost all the factors we might ever wish to use from these domains, whereas the social sciences do not. This data and models make it easy to implement these factors (i.e. we know their values, and we know what they are affected by). We must put in extra effort and take extra care to include important social and 'soft' factors in our System Dynamics models. Specifying mathematically how they interact may take a great deal of careful thought. All complex adaptive systems will have these components, and they will often play important roles in dynamics, so we cannot ignore them simply because they are more difficult to specify.

Once a System Dynamics model is built, it is 'run', this produces outputs, such as those seen in Fig. 8.2. Here we see values for the stock 'Tourists', through time, under multiple different scenarios (labelled capacity, access, and price, there is also a baseline scenario and some real data plotted). A scenario represents a change in the model, either in the equations of flows or in the structure of the model (i.e. with a connection added or removed), both of which are intended to represent some intervention or difference in the system. Here the scenarios represent different policy options. Quite different sorts of system futures are visible under different scenarios. For example, increasing tourist access or capacity ultimately leads to a crash in tourist numbers as the resulting pollution from high tourist numbers destroys the destination's attractiveness. A policy of lowering prices does not have such large impact on tourist numbers and hence does not lead to a boom-and-bust dynamic.

There is ongoing debate in the System Dynamics community, and other simulation communities, as to whether we should ever (try to) interpret model outputs as predictions of future states. Many scholars contend that as the models are simplifications of reality, and the reality we are modelling is a complex adaptive system with many components and interactions, models will never be able to provide reliable predictions of future states, so we should not attempt to use them in this way. Rather, they suggest, we should consider model outputs as hypothetical futures, or qualitative forecasts of the types of patterns and dynamics we might see. However, others take an opposing view, seeing prediction as one of the main purposes of models, and while acknowledging the difficulty of prediction in complex systems, still wish to attempt to predict using models like System Dynamics,

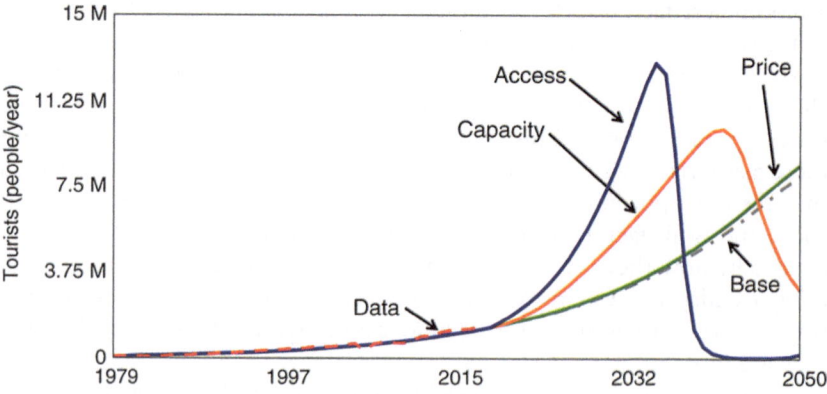

Fig. 8.2 Number of tourists under different policy scenarios. (Source: Kapmeier and Gonçalves (2018))

and (partially) assess the quality of models against their ability to predict. Without some element of prediction, they see models as missing a key element of what defines a 'good' model. We tend to sit on the fence on these debates, taking a pluralist view and seeing the merit in both approaches. As we have often said in this book, purpose drives the model. If our purpose is to predict, and we have designed with this in mind, it is not beyond the reach of methods like System Dynamics, but it is certainly very difficult, and success may well be fleeting.

As with many quantitative modelling methods, there are a range of rather technical activities that will be part of any System Dynamics process, which include parameterisation, calibration, validation, and sensitivity analysis. In practice, these elements often merge into one another, and the exact nature of how they are done, in what order and combination, and their importance in the modelling process, will depend on the purpose of each modelling project. We do not wish to go into the gory details of how they are done, which is beyond the scope of this chapter, but briefly, here are some simple (yet still debatable) definitions of them: parameterisation is the process of picking values for parameters (e.g. stocks, factors, equations, modifying flows) in the model based on data or plausible values (i.e. 'plausible' is a loose idea, but here we mean values which an expert on the system would deem sensible); calibration is the process of picking values for parameters such that the model produces outputs that reproduce some target (e.g. changing values until the model produces a pattern seen in real data); validation is the process of assessing how well the model reproduces target outputs (and often involves elements of calibration); and sensitivity analysis is where different parameters, or structural components of the model (e.g. which connections and relationships are represented), are changed systematically to assess how sensitive model outputs are to them. Each of these activities can be done in informal qualitative ways, or in much more technically demanding quantitative ways.

There are some important variations in practice and terminology in System Dynamics. In the practice of System Dynamics modelling, these mostly revolve around how we arrive at a fully specified and quantified stock and flow diagram. Some system dynamicists dive straight into a stock and flow diagram, whereas others first produce Causal Loop Diagrams (see Chap. 4) or other similar types of qualitative or semi-quantitative systems maps. These differences in where to start will likely have impacts on the models produced. Causal Loop Diagrams are a point of variation in terminology too. We have observed Causal Loop Diagrams referred to widely as 'system maps'. The way they are built appears to be consistent, but the different terminology can be confusing and unhelpful when the term 'system maps' also describes a variety of approaches, as we outline in this book. We have also seen stock and flow diagrams referred to as Causal Loop Diagrams and vice versa, though their intended use is normally clear.

Before we move on to how to do System Dynamics, we would like to emphasise one element in the method that is different to some of the others in this book but is not immediately obvious. Despite having 'system' in the name, System Dynamics models tend to be focused on specific dynamical problems rather than whole systems. In practice, this means that models rarely aim to represent a whole system, but rather focus on carefully constrained sub-sets of a system with the relevant dynamics. This nuance in focus is often not understood, with many researchers and practitioners envisaging System Dynamics as taking a whole-systems view, where in fact it is the 'dynamics' that the method emphasises. Indeed, part of the 'art' of System Dynamics modelling is finding and defining these knotty dynamical questions and problems which the method can contribute to significantly.

How Do You Do System Dynamics?

The process of conducting a System Dynamics modelling project can be broken up into six steps. These make sense conceptually, but are never actually done neatly in sequential order, rather much iteration and jumping backward and forward is the norm. They are:

1. **Problem definition**: to start, we need to pick a system and a dynamical problem or question within it. This stage is arguably the most difficult, and the one for which your approach will change and improve most as you gain experience; new system dynamicists often want to model a whole system, or struggle to find a dynamic situation where the modelling will work best. Problem definition is often done in partnership with stakeholders or users of the model and research. Taking their views on board and implementing them in the project and model is difficult, as with all methods, but particularly so with System Dynamics because it is one of the most formalised, and thus potentially rigid, of the methods in this book.

2. **Model conceptualisation**: this is the stage at which we begin to map out the dynamic problem in something that resembles a 'boxes

and connections' model. Most commonly this is done with Causal Loop Diagrams (see Chap. 4 for an introduction and description of how to use this method, we don't reproduce it here in the interests of avoiding repetition) but may also be done with other qualitative systems mapping approaches akin to Participatory Systems Mapping (see Chap. 5) or Fuzzy Cognitive Mapping (Chap. 6). The point is to use something which is easy to build but which also provides the building blocks for the quantified model of stocks and flows.

3. **Model formulation**: in this stage we take our conceptual model and turn it into a fully specified and quantified stock and flow model. The process for doing this is one of judgement and modelling choices (of which there will be many), rather than a simple conversion or standardised process. You will need to decide what elements should be stocks, which should be flows, and which should be factors that affect flows. Finally, you will need to specify the equations which determine the flows. These are often simple but can involve quite advanced mathematics if you so wish.

4. **Validation and simulation**: once the model is fully built, it is time to 'run' it (i.e. crunching the numbers, letting the stock values and flow equations play out). Normally, you do this multiple if not many times, systematically varying the model design or parameters. This is done either as part of calibration and validation, or to represent the different scenarios you might wish to explore, or as both, in many iterations. We suggest specifying a set of experiments to help you think thoroughly about different model runs you wish to do, and help you communicate this. You may require longitudinal data on the system or problem you are modelling, against which to compare your model results. This stage often takes much longer than people imagine, and uncovers issues in the model build, conceptualisation, or even in the problem definition, that require returning to those stage to fix—this is healthy iteration and should not be avoided. Your model will produce reams and reams of simulated data, normally in CSV (comma separated values) file format.

5. **Analysis**: once you have your model outputs, and any real data you are going to use, there will be a significant amount of time spent on analysing, understanding, and presenting those results. You should follow good practice in terms of planning and recording your analysis so it can be reproduced by you or others at a later stage. Equally, you are likely to want to think carefully about how you visualise

outputs and should look at similar examples you think work well, or follow more general guidance on data visualisation.

6. **Interpretation**: finally, once we have results we can share (and importantly, our descriptions and explanations of them), it is time to interpret and use these results with stakeholders and end users. Again, this is a key and often underestimated stage. The most impactful and successful projects will give plenty of time for results to 'stew' and be interpreted and allow space for further iteration from there.

The exact detail of each of these steps and how you iterate within and between them will depend on many factors, most importantly your own preferences and the purpose of the project. We cannot emphasise this enough: do not be worried about jumping back and forward, the best modellers do, and are creative as they go.

As with most of the methods in this book, System Dynamics modelling can be done in a participatory setting, and this is common. Sometimes it is referred to as 'group model building', 'mediated modelling', or 'Participatory System Dynamics'. A key decision is what stages of the modelling to do with stakeholders; do we just do the problem definition and model conceptualisation, then return with results, or do we also include them in the model formalisation stage? Both will be valid depending on your purpose. We would encourage you to attempt to include stakeholders at all stages, and not assume they do not want to get into technical details. The last thing we want in a participatory modelling project is for stakeholders to think there are 'highly technical and skilled' elements which they should not play a role in and thus passively wait for your return 'with the answers'. It may go against our instincts, but ideally, we want stakeholders to critique and attack our models, so we can uncover bad assumptions, misunderstandings, and the gaps that hamper consensus being reached. The trick is to have just enough legitimacy and respect with stakeholders so that they will take part and engage, but not so much that they won't be sceptical and criticise you.

There are many software options available for a budding system dynamicist. Point-and-click options are common, and you will not need to write the computer code for a model if you do not want to. Some of the most popular and powerful include (in alphabetical order):

- **NetLogo**: this free and open-source agent-based modelling software also has an easy-to-use System Dynamics functionality.
- **iThink and Stella**: these are two product names for a range of commercial System Dynamics software produced by isee systems.
- **Powersim Studio**: commercial software, with some free versions available.
- **Vensim**: commercial software, with some free versions available.

Though it is no longer in use, the DYNAMO simulation language is worth a special mention. This was the language used for many early System Dynamics models, including the Limits to Growth model, and is still spoken about with affection. Some efforts exist to bring it back to (some form of) life, for more, see https://github.com/bfix/dynamo.

Common Issues and 'Tricks of the Trade'

We touched on one of the more fundamental but common issues above—struggling to find a dynamical problem or question which is well-suited to a System Dynamics model. We often begin a project with a broad topic or interest, and perhaps a method in mind, and need to find a way in which the latter can be of use in the former. This is a difficult art. Some rules of thumb for defining the 'target' of the model (i.e. the thing we want to model) include:

- Avoid trying to model a whole system.
- Look for dynamical sub-parts of a system and look for issues or problems which persist despite the presence of dynamics and interventions (i.e. a problem which persists because there is no intervention or any dynamics affecting it is not really a dynamical problem).
- Ask for advice from more experienced modellers and more generally share and air your ideas regularly.
- Think about what the outputs might look like—can you sketch the graphs over time your model might produce?
- Dive into some modelling knowing you are likely to iterate back to your ideas to update them as you bump into issues building a model.

It is relatively easy to start building System Dynamics models, but much harder to produce 'good' models and develop genuine insights. This can create a common sense of frustration for intelligent but inexperienced modellers. While the technical craft can be picked up quickly, you will likely bump into many conceptual issues and questions which are more difficult to 'solve', and you will also start to see a range of issues and weaknesses in your model and the purpose you are putting it to. While as modellers we are often told not to 'fall in love' with our model, it can be just as common and problematic for new modellers to become disillusioned at the difficultly of good modelling and lose interest. This is normal and part of the learning process. This can happen with any of the methods in this book, but it is particularly relevant here because System Dynamics is a simulation method with many components to be specified, calibrated, a model to be validated and scenarios to be designed and run—in short, there are simply more points at which to run aground.

One of the keys to avoiding this issue, and many others, is to iterate quickly and aggressively. You will see calls for iteration everywhere in research and modelling, and especially for work on complex adaptive systems, where iteration is a key heuristic in researching and engaging with systems. But how do you actually do it? What does it look like? Most simply, it means jumping backwards and forward between the six steps we describe above. There is a difficult balance to doing this though, it can be done too slow and too fast. The trick is in having the confidence to move between steps openly and with purpose, but to do so in a systematic way rather than in a panic.

What Is System Dynamics Good and Bad At?

System Dynamics' real strength is in helping us develop high-level strategic insights on dynamical questions. It can help us find counter-intuitive dynamics, to rigorously play out the implications of an intervention or theory, or to play a facilitating role in doing this with stakeholders. It also helps us think about dynamics rather than individual events or static models. As well as producing outputs showing how variables change dynamically, many systems dynamics software packages allow you to see how dynamics emerge from interacting feedback loops as they change in strength and importance as a model unfolds. This breaking down of dynamics into component parts can allow more intuitive understanding of how dynamics arise for stakeholders and modellers alike. Thinking in a dynamical way is a very different way of thinking for many people, which can be hard to develop; System Dynamics will help you do it. As a simulation method, it is also well-suited to helping us test out assumptions and the potential impacts of interventions; to exploring those 'what if?' questions.

Despite being a quantitative method, it is not useful for operational or more local, micro, questions, or more generally in detailed analysis of systems. It normally represents systems at an aggregate level and focuses on dynamical subsystems. It requires a relatively large amount of time to conduct compared to other methods in this book. This can be a serious barrier to its use in applied and participatory contexts. It can also be difficult to include social and 'soft' factors in a model, where we don't often have the help of previous modelling and quantitation of these ideas to inform our implementation of them.

A Brief History of System Dynamics

The method appeared in the 1950s, developed by Jay Forrester, who used it to explore the success and failure of businesses and industrial dynamics. Legend has it, Forrester used 'hand simulations' (we can only assume these were calculations done on paper, rather than physical representations of the stocks and flows!) to explore employment dynamics with managers at General Electric. Forrester and his team quickly moved on to modelling on computers, and his approach was applied to multiple business and management questions. By the late 1960s and into the 1970s, the method started to spread beyond these roots and was famously used in the Limits to Growth work described above. Since then, it has been applied in many domains and enjoys a large and active community of practitioners and researchers.

Because of the method's name, and partly because of its approach, people often assume there are strong connections with systems science and thinking; however, the method did not emerge out of these fields, so the links can sometimes be weaker than we might expect or hope. Nonetheless, the underpinning interest in complex adaptive systems is present in both. Prominent figures in the field have suggested progress has been held back by the lack of connection to other modelling and research domains, and this has led to a lack of development in the method itself (e.g. Sterman, 2018).

There have been, and continue to be, many debates in the community. They have revolved around questions such as: whether we should aim to make point predictions; whether we have to actually build a simulation, rather than just a static model, to be 'doing' System Dynamics; and what types of validation are appropriate and how quantitative they should be. There are also different schools of thought on how the method should be used, either as a participatory tool, or as a purely analytical tool. All of these sorts of debates are common in simulation methods, though they do appear to be had rather more forcefully in this community as opposed to others. We believe there should be room for all the different ways of doing System Dynamics to exist under the one umbrella. Modelling should be driven by purpose, and as long as there are different purposes behind System Dynamics projects, there will be justifications for doing it in different ways; with or without a simulation, with stakeholders or as a highly academic endeavour; with modern quantitative validation methods, or with more qualitative and heuristic validation.

GETTING STARTED WITH SYSTEM DYNAMICS

Because System Dynamics is somewhat more formal than some of the other methods in this book, we are little more cautious than our usual 'dive in and see what happens' approach to getting started. The benefits of doing some more reading, and maybe even attending a course, are likely to be high for this method.

Perhaps, the most obvious place to start is the System Dynamics Society. The society was setup in 1983 with Forrester as president and has continues to be the main voice and organising force in the field. Its journal— System Dynamics Review—is one of the main places to find current work using the method, and its annual conference is worth considering too. Its website is easy to get lost in, has a plethora of resources, and can point you to up-to-date information on events and training, including an online course catalogue, readings, and videos.

We would recommend looking at some of the following further reading and resources:

- Loopy (https://ncase.me/loopy/) is an extremely easy-to-use and fun browser-based quasi-System Dynamics tool. Though its functionality is limited to the very basics, it is great tool to get a quick sense of how modelling dynamical systems in causal networks feels. It can be used to play around with, or to build and share simple models.
- The special issue in 2018 of the System Dynamics Review 'Celebrating the 60th anniversary of the System Dynamics field' is a good place to

start on more academic and detailed pieces. John Sterman's introduction is an excellent, if very strong, description of some of the history of the field and a call to arms to improve practice and link with other domains.

- For those who like to return to the classics of a field, Industrial Dynamics by Jay Forrester (1968), and Business Dynamics by John Sterman (Sterman, 2000) are the obvious choices, and both still considered relevant to budding and experienced system dynamicists.

Despite the need to take a little more care with System Dynamics than some of the other methods in this book, we hope you are not discouraged. You can still easily have a play with some of the software above or online models such as isee's COVID-19 simulator (https://exchange.iseesystems.com/public/isee/covid-19-simulator). Dip your toe in and see how it takes you. The joy of producing some dynamic output can really capture your imagination. It is also worth emphasising, it is fine to be self-taught in more formal methods like System Dynamics. Most researchers and practitioners are. Don't be put off by the fact you are not going to enrol on a masters or PhD programme in System Dynamics. If nothing else, giving System Dynamics a go will help you to understand what it means to 'think dynamically'. In our experience, this is a rare, difficult, and valuable skill.

References

Forrester, J. W. (1968). *Industrial dynamics*. Productivity Press.

Kapmeier, F., & Gonçalves, P. (2018). Wasted paradise? Policies for Small Island States to manage tourism-driven growth while controlling waste generation: The case of the Maldives. *System Dynamics Review, 34*(1–2), 172–221. https://doi.org/10.1002/sdr.1607

Meadows, D. H., Meadows, D. L., Randers, J., & Behrens, W. W., III. (1972). *Limits to growth. A report for the Club of Rome's project on the predicament of mankind*. Universe Books.

Sterman, J. D. (2000). *Business dynamics: Systems thinking and modeling for a complex world*. McGraw Hill.

Sterman, J. D. (2018). System Dynamics at sixty: The path forward. *System Dynamics Review, 34*(1–2), 5–47. https://doi.org/10.1002/sdr.1601

CHAPTER 9

What Data and Evidence Can You Build System Maps From?

Abstract This chapter takes a step back from individual systems mapping methods and considers what evidence and data we might use to underpin the design of system maps. It presents four broad types—data from participatory processes, qualitative data, existing evidence, and quantitative data—and outlines the pros and cons of each, considers how you can use them, and makes a call, ideally, for using them in combination and in creative ways.

Keywords Systems mapping • Information • Evidence • Data

In each of the methods chapters in this book we have considered how exactly you can use a method; what things you need to do, what you need to collect, how you create a map or model, and do analysis. We have tried to be agnostic throughout about whether the use of methods should be based on quantitative data, formal evidence, qualitative data, or participatory workshops and processes (though some bias towards participatory processes may have slipped in since this is what we are most experienced in). In practice, any of these approaches can be used, and often are combined and overlap with one another.

The fact that these types of information for building maps overlap means it is worth trying to be clear about what they are. Figure 9.1 attempts to do this in a simple way. It roughly defines the four types. We can see some of the overlaps; participatory processes can also collect

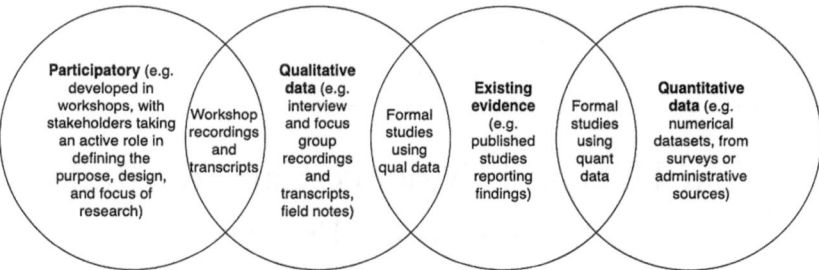

Fig. 9.1 Types of information for building system maps, and their overlaps. Source: authors' creation

qualitative data, and that evidence can be based on quantitative and/or qualitative data. Arguably, evidence can also be based on participatory processes, but it is rarely framed in this way.

This chapter takes a step back and seeks to reflect on the question of what types of data and evidence we can use to build maps, to identify to some of the key pros and cons of each, and to consider how we might go about using them. Chapter 10 goes into detail on how to run systems mapping workshops, so we won't cover that in this chapter. Rather, we first put forward a defence of the, sometimes critiqued, use of stakeholder opinion to build maps in a participatory mode. We follow this with a more practical consideration of how we can use more traditional data (qualitative and quantitative) and existing evidence to develop maps, including issues we need to take care around. We conclude with a few remarks on combining different types of data and evidence, and on the question of how to choose methods within different data availability contexts.

Defending the Use of a Participatory Process to Build and Use Your Map

One of the most common questions we get when running workshops or presenting on systems mapping is 'what evidence do you use to back up the model?' or, the closely related, 'how do you validate the model?' These are important questions, but questions that people from diverse backgrounds, and who have used different modelling types, approach in a variety of ways. Even the technical term 'validation' can have different meanings, nuances, and interpretations in different modelling domains. It is not always clear whether people think qualitative and workshop-generated data can be used to build a system map, but when asked, we often assume these types of question are based on a belief that they cannot, or that quantitative data would be preferable. The assumption appears to be that we need quantitative data or scientific evidence to trust or draw value from our maps, from our models. This reflects a strong belief held by some researchers that only quantitatively validated models, of any type, are valid or useful.

We must be upfront about a fundamental belief that many systems mapping practitioners have. That is, when working in a genuine complex adaptive system, it is highly unlikely we will have access to the breadth or depth of quantitative data or evidence we need to formally validate every node and edge in a system map, or to validate simulation outputs from something like System Dynamics. Even where we do have some data or evidence, it is likely to be patchy, with systematic reasons determining the areas we have data and evidence for, and those that we do not. The reasons for absent or patchy evidence are multiple. Many of the important components in a system are social or behavioural; they are actual practice on the ground, peoples' perceptions, tacit, and local knowledge determine what they do. Data is not generally collected on many of these things; indeed it is difficult and expensive to do so. Moreover, the amount of time needed to collect data to validate on the ground knowledge is prohibitive in contexts in which decisions need to be made with reasonable haste, that is, in policy-making contexts. Genuine complex adaptive systems are also open, causal connections may come from many different domains, and we cannot always determine in advance what is relevant. In these systems, things are also always changing; focusing on onerous data collection may make us reticent to update maps with new knowledge. Therefore, the task of validating a model based on quantitative data or evidence will always be an uphill battle, if possible at all.

This should not mean we abandon any hope of using systems mapping, or that we build narrow models that only include things we have quantitative data or evidence on. Indeed, many of the methods in this book work best, relative to other methods, in data-poor contexts. Vitally, we believe there is still huge potential value in maps that are not underpinned directly by quantitative evidence. There are many situations in which we need to make decisions quickly, on the ground, in data-poor situations, but in which we think it is likely that system interconnections are present and likely to be important. Thus, we still want to think through system interconnections more thoroughly, even without data. The pragmatic response here is to use participatory approaches to illustrate that system effects might indeed be important and to help us both raise awareness of this and start to think things through. We are likely to make better decisions with a map like this than without, so it is still worth doing. We should also note, these methods have an important function in generating new questions, for example, what sort of systems effects might be present, and might we need to take account of? How do we adjust our strategies to take account of these possibilities? What extra data do we need to gather to see what is actually happening? These are often hugely valuable and may directly inform new data collection processes.

So, rather than unfairly critique or abandon systems mapping efforts in data-poor contexts, we should proceed with the right amount of caution, and when working in a participatory mode, with stakeholders in the system, emphasise the value of their beliefs in constructing useful models. More than simply emphasising the value of participatory models, in contexts where there is long-standing acceptance and deference to the idealised standards of what science is (i.e. falsifiability, empiricism, etc), we may need to be robust and ambitious in our advocacy of these forms of information.

The main sources of value in using maps and models built in participatory ways are as discussion and thinking tools. In reality, all models ultimately have this purpose—they are there to improve the quality of our thinking and discussion—even if we become overly focused on forecasting or related pursuits (note, we do not intend to dismiss forecasting based on validated models, as an activity, but rather to make clear, it is not the only thing we can do with models).

Maps built in a participatory fashion help us to surface and explore peoples' mental models of a system or issue. By building them together, we surface assumptions and beliefs as a group that otherwise might be hidden or left undiscussed. The maps become 'boundary objects' (see Star & Griesemer, 1989) around which stakeholders and researchers can learn. They can help build consensus and capacity to make decisions around an issue, but also to find the places where disagreements are, and help us work through these and preserve ideas and represent both sides, if they are not to be resolved. Maps, and the researchers working on them, become 'interested amateurs' (Dennett, 2014; Johnson, 2015) in the system at hand; actors or objects that can be critiqued and improved by participants without the need to offend other stakeholders and their opinions. These types of use and value of participatory maps can also be developed from their analysis and outputs, not just their construction; analysis should always be seen as a participatory and iterative activity too.

In any applied situation, we need to fully understand the context and needs of stakeholders, users, and clients of our systems mapping research. These, along with our purpose and aims, will shape much of our work. A deep understanding of these contexts and needs is only likely to come from serious and iterative interaction with these people. This means, even where we are building maps from data and evidence, and perhaps are enjoying the opportunity to 'geek-out' on more formal methods, we still need to place this in a human context of use and value in which there will be some element of participation in a map's use, even if we don't really acknowledge it.

This is as far as we will go into the debate about the validity and value of participatory models. Others have covered this ground before (though these debates are often found in discipline-specific spaces, so more general lessons can be missed), for example, Voinov and Bousquet (2010), in an influential paper, outline and define core concepts for modelling with stakeholders, provide a detailed defence of drawing value from the process of modelling, rather than the results, and identify principles for modelling with stakeholders; Voinov et al. (2016) update and build on this, with a review on the topics covered in 2010; Hurlbert and Gupta (2015) consider when participation can be useful in research more generally, and what it can achieve in different contexts, many of these arguments apply to participatory modelling too; Prell et al. (2007) provide an in-depth consideration of why we might use participatory modelling, and how we can do it; finally, Voinov et al. (2018) provide a detailed assessment of the process of choosing different participatory modelling approaches. If you are going to be using systems mapping, or any modelling, in a participatory mode, we suggest you become familiar with some of this literature and the arguments in it.

Using Qualitative Data to Build Your Map

It is not uncommon to see system maps that have been built based on qualitative data. That is, data which may have been collected from interviews or focus groups, but where the map was not built during that interaction, but rather based on the recording, transcript, or analysis of it. In this approach, the textual data is converted into the boxes and connections of a systems map by the researcher. We are looking for assertions of what factors exist and how they are connected in the data. These assertions may be put quite simply (i.e. an interview says X is caused by Y) or may need to be extracted from participants' descriptions and narratives of processes and actors within a system.

One of the key advantages of using qualitative data is that it is typically both rich and broad. It can go into a high level of detail and description about a system but can also easily cover the full spectrum of relevant issues and domains in a system. This means that in theory its coverage of a system can be both broad and detailed, though probably only where we have a lot of data. A second key advantage, similar to participatory processes, is that if we are collecting the data ourselves in an interview or focus group, we can adapt our questions and prompts in ways we see fit, guiding the collection process so that it meets our mapping needs.

There are disadvantages to using qualitative data to build system maps. Foremost, it puts a lot of power and responsibility in the judgement of the researcher or modeller to translate data into a map. This is rarely a straightforward process, especially when we are using data that was already collected or was collected with additional purposes in mind. There will be many dozens of decisions about how to create factors and connections that meet the 'rules' of the method while also reflecting what was said. There is also often disagreement or contradictions in what different stakeholders have said in interviews. These will need to be resolved or preserved by the researcher. Finally, it often happens that one interviewee describes something that can give rise to one map, another interviewee gives us something else, and when we combine them we have a new map, which no individual actually described. The map we end up with is a composite that no stakeholder has described or had the chance to react to and comment on. This is not necessarily a problem, but is an important point to reflect on; is this map valid, does it reflect the mental models of our participants?

There are several software options that can help us develop system maps from qualitative data. Almost all the software we mention in this book can be used to build a map, but those which allow us to connect to or tag qualitative data for a map in useful ways are what are of value here. Below, we describe three well-used qualitative data analysis software packages, and one purpose-built application, which allow us to do this. However, there are important constraints on their functionality, particularly around exporting maps in usable data format, and none are free to use.

- **NVivo (version R1):** this version of NVivo has three types of map visualisation which have similarities to systems maps. There are simple 'mind maps' which you can create to brainstorm ideas, but which do not connect to the qualitative data or codes in your analysis. Next are 'project maps', which visualise a range of default nodes (e.g. documents, codes) and relationships between them (e.g. 'contained within', 'occurs at same time as'). Importantly, you can create custom relationships, which you could use to represent causal influence. Finally, there are 'concept maps' which are the most flexible visualisation mode in the software, allowing you to draw many types of nodes, relationships, and annotations, and connect these to your qualitative data and analysis. This is the visualisation mode in NVivo you will most likely want to use. Unfortunately, you cannot export any of these visualisations in a standard network data format (e.g. gml, json), or markup language you could use to extract the network data from (e.g. XML). This is commercial software, but many academic institutions have general licences.
- **Atlas.ti:** fundamental to the design and operation of Atlas.ti is a network structure and visualisation of your qualitative data and analysis. This works in a similar way to 'project maps' in NVivo, showing default types of nodes and relationships, with the option to manually add more. Because the network structure is central to the design of the software, and not a visualisation 'add-on', when you add relationships and nodes, these are added to all your other analysis in the software. Again, you cannot easily export the network data; however, you can export XPS files (a Microsoft Windows file type, similar to PDF), which advanced Windows users may be able to extract network data from. This is commercial software, but many academic institutions have general licences.
- **MAXQDA:** this software has a 'maps' visualisation function, which allows you to view relationships between default node types and relationships in a similar mode to those above. You could in theory create codes which represent factors in a system and connect these to generate your map, but this seems a rather clunky way of generating a system map. You cannot export the network data. This is a commercial software, but many academic institutions have general licences.
- **Causal map (https://causalmap.app/):** finally, it is worth mentioning this purpose-built software for building causal maps from

qualitative data. It is still in development, but a beta version is available. We have not used it ourselves, but it appears to have all the functionality we could want for developing, analysing, and exporting maps. It is not free, there is a free trial, but prices start at £490 a year; there is an R package forthcoming.

Using Existing Evidence to Build Your Map

Before we consider quantitative data, it is worth reflecting on how we might use existing evidence to inform a map. By 'existing evidence' we mean any study or analysis which has already been done based on any type of data. This might include peer-reviewed academic studies, or grey literature from reputable sources. We would need to define some inclusion criteria, but we would expect that it would include similar sources to those which make it into systematic reviews, meta-analyses, or rapid evidence reviews. In this mode, we are in effect doing a systematic review or evidence review focused on building a causal description of a system.

One of the main advantages of doing this will be that we have strong evidential support for the map we create. Depending on how strict the inclusion criteria we set are, we will be able to make pretty strong assertions about the validity of our map if it is backed up by peer-reviewed studies. Given the formalism of the studies and analysis we will likely use, and the pervasive use of concepts such as variables and causal relationships, there will also be less of a need for a modeller or researcher to make lots of decisions in translating individual pieces of evidence into nodes and edges in a map; this conversion should be more straightforward than the equivalent process for qualitative data.

However, there will still be some researcher judgement involved, and again the issue of combining pieces of evidence which perhaps contradict each other. More important though, will be the issue of coverage of a system. In many systems, there simply will not be much relevant evidence we can use. Where there is evidence there will likely be systematic reasons for what is covered and what is not, with corresponding risks of systematic bias in the resulting map. We will need to be careful about areas of a system that are not covered by existing studies, and how we can account for them. This means this approach is only really applicable in domains where there is a strong tradition of evidence gathering in formal studies of all aspects of the system, or where we are only hoping to use evidence to inform part of a map, not all of it.

The process of using evidence in this way will start with doing the early stages of a more traditional evidence review or systematic review. We recommend you look at some of the guidance for using these approaches in the domains in which you are working and think about how you could use them to gather and classify evidence in your system. For systematic reviews, Petticrew and Roberts (2005) is accessible and thorough; for rapid evidence reviews, you may find Crawford et al. (2015) useful, though it is applied to health care, so may need translating for other contexts; finally, for evidence mapping, try O'Leary et al. (2017) for inspiration. You will likely not need to complete the full process for these approaches, but just do sufficient to collect and classify the evidence relevant to you. From here, the process will involve translating each piece of evidence into the nodes and edges it can underpin and adding these to your map. As you progress it will likely be useful to have a version of your map which is annotated in some way to record and visualise what evidence is underpinning what bits. You might also want to use this in the analysis of the map, for example, to choose your focus (perhaps on bits with less evidence).

Using Quantitative Data to Build Your Map

There are two different ways to use quantitative data directly in building your map. Both rely on the idea that the variables we have in a dataset are appropriate to use directly as the boxes in a system map. This is not a silly assumption but should be checked and thought through—you may want to consider whether the variables are at similar scales and whether there are likely to be lots of important factors not present in your data.

In the first mode of using quantitative data, you may have data of any quantitative form (e.g. time series with data points for multiple variables through time, or cross-sectional with only data at only one time point) that allows you to do analysis on the statistical and/or causal relationship(s) between variables. This will likely use traditional statistical approaches (e.g. regression) or causal inference methods (e.g. difference-in-differences—see Cunningham, 2021, for an introduction) to focus on which variables we might find an association or causal relationship between. When you find a relationship between variables, you can add a connection between them. Depending on the method you use, you may be able to use the analysis to tell you the direction of the arrow, but in many cases, you will have to use other knowledge or theory, or leave the connection undirected. You will need to decide and use a threshold at which point a relationship is 'significant' enough to draw a connection between two variables.

In a second mode, if you have time series data across a wide range of variables you think may be relevant in your system, you can analyse these as a whole to extract causal networks directly using one or several of a variety of methods for 'estimating networks'. These methods rely on a range of measures, such as 'conditional independence'. This is an important but tricky concept; take, for example, two variables, a child's height and the number of words they know. These may appear to be related, but it is actually the child's age which has a strong influence on them both; they are not causally related. So, a simple two-variable analysis may make them appear related, but if we include age, we will conclude they are 'conditionally independent'. This process of looking at two variables, considering others, is useful for constructing networks from data. This mode of using quantitative data is less widespread, and still in its infancy in the social sciences, but has the potential to be powerful and quick to use.

The main advantage of using quantitative data in either of these ways is that we have a direct and fully transparent connection between the map and the information we have based it on. At the 'information-to-map' stage of the process, there is the least amount of researcher judgement required compared to other modes. We have conducted the analysis, perhaps collected the data, and do not have to make much, if any, interpretation to convert quantitative data to nodes and edges. Nodes will be the variables we have in our data, and connections will be drawn where our analysis meets some threshold (likely defined using existing norms and standards) in suggesting some association or causal relationship. However, we must not forget that a lot of researcher judgement is involved earlier in the process when we curate datasets and choose methods. Quantitative datasets are often treated as objective truth, but we must always keep in mind they are constructed and collected based on judgements, can contain errors, and offer only a snapshot of some part of a system.

The most salient disadvantages, or risks, to using quantitative data to directly inform maps are similar to those for existing evidence. It restricts our map to only those things which are both quantifiable and which we have data on. As we have repeatedly stated, this is often a serious restriction; we do not typically have a wealth of (good or even usable) data on a system, and where we do have it, there will be biases in what it includes and excludes. When using quantitative data, we must have ways to account for the aspects of a system for which data is difficult or impossible to collect and use. We should also be aware of the choices we make in what system to look at, when these are driven, at least in part, by what data we think is available. We must also be careful with our language and describe the arrows in our map using the language that the statistical or causal inference method uses. So, for example, if we use traditional statistical method, which does not allow us to talk about cause, we should only describe connections as showing relationships or associations, rather than causes.

Health warnings noted, let's think about how you might actually use quantitative data. To underpin the inclusion of individual nodes and connections, you will first need to get hold of the data, and then use one, or several, of the many statistical and causal inference methods that exist. There are dozens, if not hundreds, of these methods (too many to mention here). They can be grouped in different ways, from those that only assert association, such as standard linear regression, through to causal inference methods such as difference-in-differences and instrumental variables (again, see Cunningham, 2021, for an introduction). Methods can also be grouped by the types of data they can be used on. If you have cross-sectional data, then 'structural equation models' are one of the most obvious approaches to use; there is a range of methods within this broad umbrella, for an introduction we recommend the introductory materials on the UK National Centre for Research Methods website—https://www.ncrm.ac.uk/resources/online/all/?id=10416. If you have time series data, then 'Granger causality' approaches are one of the most popular (Granger, 1980). Finally, if you have panel data (i.e. longitudinal data where the observations are from the same subjects in each time period), you can use causal inference methods such as difference-in-differences.

It is not within the remit of this book, nor do we have space, to introduce these methods. If you are not familiar with them already, we strongly recommend you develop a basic understanding of the range of methods, their pros and cons, and then dive deeper into the methods you think you might want to use. To help you start on this learning journey, we recommend the following: Cunningham (2021) for an accessible introduction, Pearl et al. (2016) for a more technical but short introduction, and Peters et al. (2017) for a more traditional textbook. Once you have the data and the method(s), it is a simple task to create your map (i.e. draw edges between nodes) when you find casual relationships. You can do this step-by-step on each two-variable set you have, slowly building the map up.

If you are lucky enough to have lots of time series data relevant to your system, there are some emerging methods which make the process more streamlined. These methods are referred to with different terms, but most common are 'network estimation' or 'causal discovery'. The first step, again, will be getting hold of data. You will need to collect as much time series data as possible for your system. You should aim to collect data on as many domains as possible and look to maximise the number of time periods covered by the data. The methods themselves include basic correlation thresholding approaches, Granger causality approaches, statistical structure learning approaches including causal graphical models (i.e. Bayesian networks) and structural causal models, inner composition alignment, and cross-convergence mapping, to name a few. They all have different pros and cons revolving around what data they can use, what they can assert about causality, and what types of maps they give you (i.e. directed or not, weighted or not, cyclic or acyclic). Ospina-Forero et al. (2020) provide an excellent overview of these and other methods in the context of the sustainable development goals (SDGs). SDGs are an interesting way to frame systems because the scale of measurement (typically national) and good data collection around them mean they are often amenable to this type of analysis. Systems which are appropriate to define at a national level, or other scales at which there is lots of data collected and available, may be one of the most appropriate to develop quantitative data-driven system maps because of this data availability. Most of these methods will produce maps directly for you. The 'final' task thus becomes combining these and/or converting them into a form compliant with whichever systems mapping method you are hoping to use.

Using Different Types of Data and Evidence in Practice

Though we have outlined the use of different types of data and evidence individually in turn, we will often want to combine them. Indeed, given enough time and resources, it is hard to think of a reason why we would not want to combine them wherever possible. In a similar fashion to the combination of methods, as described in Chap. 11, combining data sources and evidence will help us approach a topic more holistically and give us helpful points around which to cross-compare, triangulate, and iterate. Different skills are needed to use different types of information, and this should not be underestimated as a barrier. Using quantitative data is technically demanding, whereas running participatory processes requires strong facilitation and communication skills.

Perhaps the most obvious combination of types of information is to develop a map in participatory mode, and then refine it with further information from different data sources and existing evidence (as is sometimes done with Bayesian Belief Networks). These could be used to annotate a map, validate it, or to inform quantification decisions (i.e. what the conditional probabilities are in a Bayesian Belief Network, or what the equations are in a System Dynamics model). The second use we see as holding most value is to use qualitative data and participatory processes to address the potential gaps in quantitative data and existing evidence. This replicates the logic of much mixed methods research in using the strengths of approaches to covers the weaknesses of others. Combining sources in any way will involve additional work in map visualisation and may create parallel streams in your analysis and use of maps.

As we often do, we want to finish with a plea for creativity. It is tempting to think of different types of data working better with certain methods. For example, you might think quantitative data will work best with Bayesian Belief Networks or System Dynamics because of their quantitative and more formal approach. However, we believe it is important to keep an open mind here; any of the methods can, and have, been used with different data sources, evidence, and processes. Moreover, some of the room for innovation, and potential new insights, may be in more unexpected and creative combinations. It may be interesting to ask, what might a Rich Picture of quantitative data look like? Or, how might a Theory of Change based on evidence look different to a Theory of Change from the mental models of a policy team? Any combination of method and information source is valid, as long as we are aware of the potential gaps and omissions in the information we use, and either address them directly, or adjust our aims and claims accordingly.

References

Crawford, C., Boyd, C., Jain, S., Khorsan, R., & Jonas, W. (2015). Rapid Evidence Assessment of the Literature (REAL©): Streamlining the systematic review process and creating utility for evidence-based health care. *BMC Research Notes, 8*(1), 631–640.

Cunningham, S. (2021). *Causal inference: The mixtape.* Yale University Press.

Dennett, D. (2014). *Intuition pumps and other tools for thinking*. Penguin.

Granger, C. (1980). Testing for causality: A personal viewpoint. *Journal of Economic Dynamics and Control, 2,* 329–352.

Hurlbert, M., & Gupta, J. (2015). The split ladder of participation: A diagnostic, strategic, and evaluation tool to assess when participation is necessary. *Environmental Science and Policy., 50,* 100–113. https://doi.org/10.1016/j. envsci.2015.01.011

Johnson, P. (2015). Agent-based models as 'interested amateurs'. *Land, 4,* 281–299.

O'Leary, B. C., Woodcock, P., Kaiser, M. J., & Pullin, A. S. (2017). Evidence maps and evidence gaps: Evidence review mapping as a method for collating and appraising evidence reviews to inform research and policy. *Environmental Evidence, 6*(1), 1–9.

Ospina-Forero, L., Castañeda, G., & Guerrero, O. A. (2020). Estimating networks of sustainable development goals. *Information & Management, 103342.* https://doi.org/10.1016/j.im.2020.103342

Pearl, J., Glymour, M., & Nicholas, P. J. (2016). *Causal inference in statistics: A primer.* Wiley.

Peters, J., Janzing, D., & Schölkopf, B. (2017). *Elements in causal inference: Foundations and learning algorithms.* The MIT Press.

Petticrew, M., & Roberts, H. (2005). *Systematic reviews in the social sciences: A practical guide.* John Wiley & Sons.

Prell, C., Hubacek, K., Reed, M., Quinn, C., Jin, N., Holden, J., Burt, T., Kirby, M., & Sendzimir, J. (2007). If you have a hammer everything looks like a nail: Traditional versus participatory model building. *Interdisciplinary Science Reviews, 32*(3), 263–282. https://doi.org/10.1179/030801807X211720

Star, S. L., & Griesemer, J. (1989). Institutional ecology, 'Translations', and Boundary objects: Amateurs and professionals on Berkeley's museum of vertebrate zoology. *Social Studies of Science, 19,* 387–420.

Voinov, A., & Bousquet, F. (2010). Modelling with stakeholders. *Environmental Modelling & Software, 25*(11), 1268–1281. https://doi.org/10.1016/j. envsoft.2010.03.007

Voinov, A., Kolagani, N., McCall, M. K., Glynn, P. D., Kragt, M. E., Ostermann, F. O., Pierce, S. A., & Ramu, P. (2016). Modelling with stakeholders—Next generation. In *Environmental modelling and software* (Vol. 77, pp. 196–220). Elsevier Ltd. https://doi.org/10.1016/j.envsoft.2015.11.016

Voinov, A., Jenni, K., Gray, S., Kolagani, N., Glynn, P. D., Bommel, P., Prell, C., Zellner, M., Paolisso, M., Jordan, R., Sterling, E., Schmitt Olabisi, L., Giabbanelli, P. J., Sun, Z., Le Page, C., Elsawah, S., BenDor, T. K., Hubacek, K., Laursen, B. K., … Smajgl, A. (2018). Tools and methods in participatory modeling: Selecting the right tool for the job. *Environmental Modelling & Software, 109,* 232–255. https://doi.org/10.1016/j.envsoft.2018.08.028

CHAPTER 10

Running Systems Mapping Workshops

Abstract This chapter explores the practicalities of organising and facilitating systems mapping workshops. It considers how you should plan them; how to choose venues, materials, and technology; how to record them; and how to improve your facilitation skills. It provides readers with guidance for dealing with some common issues, such as power dynamics, disagreement, confusion, disengagement, and facilitator burnout, and points to useful resources for specific methods.

Keywords Systems mapping • Workshops • Stakeholders • Facilitation

Mapping workshops are key milestones in any systems mapping project. In participatory projects they are vital spaces for engagement, discussion, and the creation of maps. Workshops will inform and energise the entire project. They are moments around which participatory projects can live or die. Everyone involved, including participants, will want them to be a success. This means they feel like high-pressure moments for researchers. Organising and facilitating workshops is also a skill in its own right; it is difficult to do. It takes thought and practice before you can expect to run workshops well, and to not feel exhausted by the process.

In this chapter, we will try to shed some light on how to plan and run successful workshops. Much of the craft of running workshops relies on tacit knowledge and subtle inter-personal and communication skills, so we

© The Author(s) 2022
P. Barbrook-Johnson, A. S. Penn, *Systems Mapping*,
https://doi.org/10.1007/978-3-031-01919-7_10

can't cover everything. But we will be doggedly practical and reflect on our own experiences. The style of this chapter is a little different to others; there are so many things to think about that we have used a more bullet-point-heavy structure to help you use this chapter like a checklist in your planning.

It is worth reminding ourselves that individual workshops are always part of a wider project process. You can, and should, take care in designing this process. Your purpose, and the needs of users and stakeholders, will drive the initial process design, and the process may need to change as the project develops. Considerations around process design for each method are discussed in individual method chapters, and we have developed our own practical guide for Participatory Systems Mapping processes in Penn and Barbrook-Johnson (2022).

PLANNING WORKSHOPS

It is possible to over-plan and over-think workshops and to stress yourself and others needlessly. A lot of thinking and adjusting can be done during the workshop (especially as you become more experienced), so try to avoid the urge to plan everything in microscopic detail. That said, it is vital to have a plan for workshops and the wider process they fit into, and to develop this with users and stakeholders. Key things to plan are:

- **Purpose**: you will want to have a clear idea (and written statement) of the purpose of the workshop; what the topics to be covered are, and what any outputs you hope to produce from it are. If you have multiple workshops, it should be clear how these fit together, and why they are needed.
- **Who to invite**: you will need to decide who to invite and how to invite them. Assuming there is not a specific reason to only focus on one group of people, you want to get a good spread of people, representing different views and knowledge of different parts of a system. You also want people to accept your invite, which might demand a lot of their time. So, you need to be compelling in your invite, making clear the value of your work to them.
- **When to hold it**: timing can be key to the success of a workshop when there are other processes and events unfolding relevant to your work. Avoid running your workshops at the same time as important events for your stakeholders, or during times of the year when they

will be busy. If there are other relevant processes, events, or publications being launched, think about how you would like to relate to them; should you go before them and try to inform them, or do you want to go after and use them as inputs?

- **Have a time-plan document**: the key document you will create for planning will be a time-plan of the workshop. We recommend having a detailed but flexible time-plan. If you have a three-hour session, you probably want to know what you are doing in each fifteen-minute section (sections may last longer than fifteen minutes but use it as your 'unit of time-planning'), if you have a full day, then planning in thirty- or sixty-minute chunks is sufficient. The point is not to have a rigid plan you must stick to but rather to have thought through all the things you want to do and to be realistic about what can fit in your time. A written plan is an invaluable resource during a session, to check progress, adjust thinking, and ensure you get done what you need to. We normally combine this time-plan with all the prompts and questions you plan on using.

Venue, Materials, and Technology for Workshops

Having the right venue, materials, and technology can unlock all sorts of potential problems and transform a low-energy workshop into a major success. Things to consider include:

- **The venue**: where is most convenient and familiar for your participants to meet you? Go to them, if possible, rather than asking them to come to you. As long as it does not create a perception of bias in the process for some people, it may be ideal for one of the participants to host you. You should aim for the nicest room you can get hold of, but which is still appropriate for the type and numbers of participants. Windows, good light, and fresh air are a must. Having space to move is also important, but you don't really want to be in a large hall if there are only ten of you.
- **Tables and chairs**: it is easy to forget about tables and chairs, but you may have some clear constraints on what will work. Do you want 'lecture style' (i.e. rows of seating), 'cabaret style' (i.e. multiple small tables), or 'boardroom style' (i.e. one large table), or something different? If you want groups of people to be huddled around a table drawing and discussing a map, you will need table(s) that are the

right size, and chairs which can be moved aside easily. Beware many modern office tables have holes for cables, or power sockets built it, these can easily get in the way, when all you want is a good-sized flat surface. Discuss these needs with your venue ahead of time and make sure you have time during the day to set up the room if needed.

- **Walls**: using walls to put up notepads and post-it notes is useful. Be clear on what the walls are like in your venue; can they accommodate being used in this way, or are they covered in pictures or notice-boards already, or are they antique wood which you are not allowed to touch? If you can't use walls, you may want to take some flipcharts.

- **Materials**: if you are going to be physically drawing your map with pens and paper, you need to think about these and bring them with you. You will need to have enough pens, in the range of colours you want, and the right size/thickness for the size of group (i.e. large marker pens may be cumbersome for a group of five, but using standard biro pens will mean writing is unreadable if you are in a group of twelve). Post-its are invaluable for brainstorming and for building maps which can be adjusted as you go. So too is using whiteboard paper (i.e. shiny paper which you can draw on with whiteboard pens, and then erase easily when needed with a tissue). If you are writing straight onto large pieces of standard paper, it is difficult to change a map, and people can be hesitant about writing things down which they know cannot be changed easily or erased.

- **Technology**: do you need a projector, or screens during the session? If you are just giving a quick introductory presentation, a simple setup will suffice, but if you are planning on creating your maps directly in digital form, you will need a high-resolution projector, or good Wi-Fi so people can work on their own laptops, on the same map in real time. It can be tempting to use high-tech solutions for drawing your map, but we tend to err on the side of low-tech options. There is less risk of things not working, and it is more inclusive for people who are not confident with computers (in our experience, people routinely underestimate this issue, and exclude important voices by opting for high-tech mapping sessions). It is also easier to create a sense of energy and break down barriers by having people standing together around and interacting with a physical object.

- **Existing versions of maps**: if a workshop is your second or third, and you already have a map started, or you have created a map prior to your first workshop, you will want to bring this with you. We tend to digitise maps (i.e. enter them into our software of choice for that project) and then bring large printouts of these (A1 or A0 paper size). You can then overlay a large piece of transparent whiteboard paper and continue map building with pens and post-its. Having the actual original paper and post-its to hand can be useful too, to refer to, but we don't tend to use this to continue map development directly as paper gets tatty and things inevitably move and fall off.

FACILITATING DISCUSSION

The actual act of facilitating discussion between people is based on a range of tacit and subtle inter-personal skills. You will have many of these skills already, but they will take time and practice to further develop and refine for the context of mapping workshops. On the upside, these skills are easily transferable; if you have facilitated other types of workshops, or conducted interviews, then these skills will stand you in good stead for systems mapping workshops. Once you have run a few workshops, you will feel more comfortable with other types of facilitation and public speaking. You can improve your facilitation by:

- **Having a clear plan**: as we mentioned above, have a document which outlines the plan for the workshop and includes all the information you want to have to hand. More than this, try to develop a clear plan for your approach to facilitation. Make decisions on key points ahead of time, for example: any red-lines you have on what should be included or not in the map, will you encourage others to draw or do you want to do it; will you try to be a dominant and energetic facilitator or more passive? However, you may have to adjust your approach to this depending on the dynamics of the group.
- **Having clear prompts**: a key part of any plan for facilitation will be to develop your prompts and questions and write these down. You want to have both the initial questions you will ask and the prompts you will use to encourage people to elaborate and prompts to use if discussion is slow. Having a plan for different eventualities, and some notes on these, will help you feel more relaxed; you probably won't even need to use them, but knowing they are there can be reassuring.
- **Knowing the 'rules of the method'**: the last thing you want to be doing during a workshop is worrying about what exactly the 'rules of the game' are for the method you are using. You should have

these clear in your mind beforehand so you can answer any questions about the method, but more importantly so you can focus on facilitation rather than technicalities.

- **Practising**: organising mini-workshops with colleagues, friends, or family to practise using the method and facilitating discussion can be a useful way to learn what works and how you react in a workshop setting. We recommend always doing this beforehand if you are using a method for the first time. Make it fun, choose a topic that makes sense for the people you convince to turn up, and ask them for constructive feedback.

- **Managing group dynamics**: You will have to be alert to the way the group dynamics and discussion are unfolding and adapt your facilitation and input to respond to this. Initially, there is often hesitance to overcome, you may have to lead more strongly at this stage. Your aim should be to energise and make participants feel comfortable, so that you are steering rather forcing. There are some key things to be alert to as the process develops: are one or a few people dominating discussion whilst others hang back? Work to moderate this and bring everyone into the discussion. Be alert to how comfortable people seem in the group context and whether they have something to say, but don't feel bold enough. Are people drawing connections without discussing with the rest of the group? Ensure that all proposed links are discussed and agreed. Are people getting over-focused on one specific area of the map to the exclusion of others?

- **Giving away power**: it is easy as a facilitator to feel responsibility for every moment of a workshop, and to try to control things too tightly. We recommended erring on the side of giving away power and responsibility to participants. Ask them to draw instead of you, ask them to ask questions of others, and critique the map as it emerges. Feel free to describe yourself as a non-expert on the system you are mapping; in a participatory process, positioning yourself as the expert will inhibit discussion. In an ideal situation, if participants are clear on the purpose of the session, and understand the method, by the second half of a workshop they could maybe even run the session without you!

- **Taking breaks**: facilitating is an exhausting activity. Between the nervous energy and the need to be focused, you will use a lot of physical and mental energy. Take breaks, both to give yourself a few moments rest, but also to reflect on how the session is going and adjust if needed. Participants will need breaks too. Have a rough idea of when breaks will be, but allow a little bit of flex so that you don't stop at very productive moments, or feel the need to go on when energy is low or you have reached a natural breakpoint.

- **Working in pairs**: working in pairs can be a useful way to make facilitation easier. For any given section of the workshop, you can assign roles, one person can be in active facilitation mode, and the other can be in a participant observer role, watching and reflecting, but also contributing when needed. You can then switch roles during different sections. Reflecting on how things are going, and managing energy levels, become much easier with someone else too. However, beware of working with too many people. We have found facilitation can get a bit muddled if there are maybe four or five people who participants perceive as facilitators. Side conversations can pop up, which we normally hope to avoid, or people can go 'off-message' if the purpose is not simple and clear.

CAPTURING AND RECORDING WORKSHOPS

It is vital, but difficult, to capture the discussion during a workshop. The map you create itself is not the only output. The discussions are equally valuable, and it can be useful to refer to these later. You have two basic options for capturing a session: you can either take an audio (and maybe even video) recording, and then transcribe or make notes re-watching it; or you can take written notes. We normally opt for the latter and have a dedicated note-taker present for the workshop. We ask them to take as detailed notes as possible, and then we look at these soon after the session and write them up into something more formal, which we can refer to in the future. If you wait more than 24 hours to write up notes, we find it is increasingly difficult to remember the nuances of the discussions. We also take photos of, and keep, all the flipchart notes made, and post-its which don't make it into the final map.

Using an audio recording removes the needs for a note-taker and gives you a more accurate record of the session, but creates two potential issues. First, you may inhibit conversation if people feel they are 'on the record'. Second, you will have a significant task in converting the recording into a transcript or notes. Transcription can easily take five or six times the length of the recording when there are multiple people involved.

Lastly, if you are physically drawing your map, we highly recommend you take photos of the map throughout the process. Photos of the map at the end will be vital to help you digitise (i.e. input into whatever software you want to use), but photos during will also help you reflect on the process, how the map developed, and maybe even jog your memory of the discussion.

POST-WORKSHOP

After a workshop you may just want to collapse in a heap and have a sleep! You should take some time to rest if the process is tiring, which it often is. However, it is also important to take advantage of the energy and raised interest a workshop will often generate. As well as processing notes and photos, and starting the process of digitising maps, you should contact participants to thank them for their time, explain what will happen next, and invite them for bilateral conversations should they want them. We have found that participants are often interested in following up one-to-one, and these can be some of the most useful spaces to develop ideas for analysis and use of maps. A de-brief with your team will also likely be useful; make some dedicated time to reflect on what worked and what didn't, how you might change future workshops, or the project design.

COMMON ISSUES

Workshops are not always all sweetness and light, there can be thorny issues you will need to resolve. The most important we have come across are the following:

- **Difficult power dynamics**: there are all sorts of group dynamics that can emerge during a workshop, or that can exist from previous interactions. You may not be aware of many of these, and they rarely cause serious issues. However, when there are uneven power dynamics in a group, it can undermine a workshop. You may find that the most senior or powerful person in a group dominates, and the mapping just reflects their views; or you may find that even when a powerful person does not want to dominate, others are still hesitant to say much in front of them. In either situation, you need to find ways to ensure the people who are less powerful, or more hesitant, can contribute. You may have to irritate the more powerful people to do this, but it is likely worth the risk. It can be helpful to position your-self as an outsider, or to justify your request using the method, rather than directly saying someone is dominating too much. If there are foreseeable issues here, with regards to say employees not likely to talk freely in front of a boss, or groups who have conflicted relation-ships then it can be useful to organise separate workshops for differ-ent people or groups.

- **Disagreement between stakeholders**: potentially worse than uneven group dynamics is strong disagreement between stakehold-ers. We are referring here to disagreement that is severe enough that it could lead to some stakeholders refusing to attend workshops, attending but not contributing or bringing a negative energy, or to heated arguments between individuals. If difference of opinion is very wide, and thus relations between stakeholders are poor, you may want to think about running separate workshops for them. Unless your role is specifically on reconciliation, we would avoid put-ting the extra burden on your workshop of playing this role as well as mapping. If it feels appropriate to try to get people in the same workshop, you will need to have plans and prompts to help them constructively discuss disagreements, look for possible resolutions, and/or have ways of preserving different views in the map that is created. It can be useful to emphasise the role of mapping in bring-ing our mental models and assumptions to the surface, and that the whole point is to find out where we agree and disagree, and why exactly this is. Using a map to do this can help to diffuse any anger or emotion in a discussion. Where calm discussion does not lead to agreement, you should be able to preserve two or more views of an issue in one map; this is easier in the more flexible methods. Although in general, coming to a shared understanding captured in a single map is the goal, the aims of your process should dictate your decision around whether to create more than one map version, for example, if explicitly comparing different system understandings will help illu-minate what is going on.
- **Confusion about the method**: be prepared to explain how the mapping method you are using works in two or three different ways. Think about how you would describe it to a ten-year-old child. If you can't do this, then you probably can't explain it clearly to diverse sets of stakeholders. People will misunderstand it or won't listen carefully to your first introduction of the method. Sometimes people

will think they can bend the 'rules' of the method if they want to. They may view the purpose of the session simply to have a discussion, and not see as much value in the actual map you are producing. Misunderstandings can go unnoticed until quite far into a workshop. You need to be patient with people, avoid positioning yourself as the expert and them as the lay people. It can be useful to use prompts and summaries as people are adding things to the map to ensure understanding is correct, rather than giving long presentations at the start, or giving handouts with the 'rules'. For example, if someone explains, 'I think X should be connected to Y with a negative arrow', you can prompt and summarise as it is drawn down by saying, 'OK, so that is because as X increases Y tends to go down?'. If that person actually meant 'X is bad for Y', this will quickly be realised, and you can explain again without the sense of you being patronising or needlessly concerned with the details.

- **Disengagement of stakeholders**: it is inevitable that some people will be disengaged during a workshop for one reason or another. Typically, it is only one or two people, but you can end up with half the participants not really contributing, simply because they are bored, don't see the value, or are distracted. You should not be disheartened if this happens. It will happen in the most important of processes and to the most skilled of facilitators. What you need, again, is a plan for how you will deal with it. Setting some gentle ground rules can help. For example, if people have their laptops with them and start to do other work on them, it can save a lot of effort later to ask them politely to not use their laptops during the workshop. Remember, they may have assumed they were attending a presentation, and were planning on sitting at the back, responding to emails. The trick is not to shame people who are disengaged, but to pre-empt potential reasons for them becoming disengaged, or to gently bring them back in. Creating an engaging environment, for example, asking people to stand rather than sit and having the whole table covered with mapping materials without room for laptops is often helpful.

- **Facilitator burnout**: it is likely you will experience some aspect of facilitator burnout during a project with multiple workshops. Sometimes this can come in the form of simple tiredness, but more often we find it manifests itself in more subtle ways. You might start to feel frustrated with a workshop, or set of workshops, or lose confidence in the purpose or quality of the work. You might find you start to develop nerves before and during sessions, when previously you did not, or you may begin to feel unreasonably anxious about workshops. This is totally normal. Sometimes it is a good sign you are being reflexive and thoughtful about your work. The simplest way to deal with this is, of course, to give yourself some time not working on, or thinking about, the project. A little time and reflection can work wonders. You may also want to talk with team members about how you are feeling; they may be feeling the same way, and/or can offer reassurance.

Running Online Workshops

As we write this, in late 2021 in the UK, we have spent the last year and a half largely unable to run workshops in person due to the pandemic. No one knows with any certainty what will happen in the future, if and when meeting in groups in person for workshops will become the norm again. Though in 'normal' circumstances we would almost never want to run a workshop virtually, given the changing norms brought on by the pandemic, it is worth considering how we they can be facilitated online. We have spent the last eighteen months doing this with mixed success, here are some of our reflections and tips:

- **Attendance is easier online**: interest for, and attendance at, events has stayed broadly the same. If anything, it has been easier to get the most ideal participants, as attendance takes less time and effort without travel. You might want to use this to your advantage; now that people are more used to attending events online, and if you want to reach people who are geographically far apart, it may be worth running an event online even if it is not strictly necessary.
- **Interaction and discussion are far harder to generate**: there is no doubt that discussion and interaction is poorer online. Without body language cues, and with the short delay between speaking and hearing, fast-paced discussion between three or more people is almost impossible. You will likely need to have to play a more active role facilitating people to help unlock some of these difficulties. You may want to devise a system for people to make comments, for example, using 'raise hand' functions, with you chairing actively. You can make use of a chat function too, but this can be distracting, and create

parallel discussions. It is useful to increase the facilitator to participant ratio here, so that, for example, someone is monitoring the chat, another person leading the discussion, and another keeping an overview of the interface you are using. Group size should also be reduced to help reduce these problems. You may find it useful to make use of breakout rooms for certain discussion-heavy parts of a workshop whilst keeping other parts in the whole group. All of this creates more work, for example, in creating multiple versions of a map which must be discussed and merged.

- **Shorter but more sessions can work well**: because online discussions are more difficult, people will get tired and frustrated more quickly. We have found running shorter sessions, but more of them, or following them up with small groups or one-to-one discussions can work well. For example, you might replace a four-hour in-person session with a two-hour virtual workshop, and then half- or one-hour follow-up calls with key participants. Beware, this may end up taking up more time for you. You can also solicit more information in advance of workshops. For example, using online voting or whiteboard tools to allow participants to suggest and vote on important factors to be included in a map.
- **Software options are important**: you have two key choices here. First, which software to use to handle the actual call; you need to think about which services people can access and what functionality they have you might want to use. It is best to assume people will have serious constraints on what software they can use, and to ask them directly before making your choice. Second, you will need to decide how you are going to build your map and what software to use. The simplest thing to do is to share your screen, and you build the map. However, this might be rather dull for everyone else to watch, and will likely inhibit some of the discussion, giving you too much power as judge, jury, and executioner on all mapping decisions. Alternatively, many mapping software options can be used in web browsers and have multiple people working on them at the same time. This works well when there is little lag in the changes they make. Beware of people just working away in silence though; we have found people will just start mapping on their own in different sections of the 'virtual table'. Again, assume people will have technical issues, and get these ironed out beforehand—make sure they can access the software and can use it. Some people will not be confident with it, and

you should make every effort to accommodate them, rather than excluding them. Ideally, build in a 'tech-ice-breaker' session at the start of a first workshop in which people get to use the software in an introductory activity.

- **Test software properly**: it is vital to have a proper play around with any software you are thinking of using, even software you use for other purposes, or offline. You will only discover and understand the full functionality by spending some time with the software in a non-pressure situation.

- **Physical and online mapping create different experiences and maps**: be aware of how software and hardware can subtly affect processes. For example, a full view of a systems map may not be visible on a computer screen meaning that people naturally zoom in and focus on subsections and miss the bigger picture. You may want to adapt process design to overcome this. For example, setting aside time to 'journey around' the whole map.

- **Burnout is more likely, but also more subtle**: if you adjust the structure and length of workshops for the online context, they will not be as generally tiring as an in-person one. However, they do use up a lot of mental and emotional energy, which can creep up on you in unexpected ways. Anyone who regularly holds work video calls knows that they can sometimes be slow, awkward, and even painful to be part of. Virtual workshops can be the same; as the facilitator, you will feel this the most. We have found it is normal to feel more anxious about a workshop and the overall process, when doing them online. You should expect this, build in more breaks that you would for an in-person workshop, and give yourself screen breaks before and after a workshop.

As the world has adapted to working online, there have been several useful discussions and guides developed for running workshops online, you may find the following helpful: Fowler (2020) gives some excellent broad-level suggestions; Khuri and Reed (2020) go into more detail on how to generate more interaction, and consider some of the general-purpose software tools available, some of which include functionalities like qualitative systems mapping; finally, Dialogue Matters (2020) provide more formal and detailed guidance, though technically it is targeted at online meetings rather than workshops, it provides another thorough list of considerations and tips.

Getting Started Yourself

Embarking on a workshop or series of workshops can be intimidating the first time you are tasked with it. There are many resources available for general guidance on facilitation and running workshops; numerous books on the subject, and many blogs and web-guides. But to be honest, we have used few of these, and don't have any we would particularly recommend. Have a search yourself and see if there are any produced by organisations or authors working in domains relevant for you. We have learnt the most, and developed the skills to run workshops, from attending workshops ourselves, from discussing with colleagues with more experience, and through practice, both in 'test-workshops' with colleagues, friends, or family, and in the real thing.

We recommend organising a practice workshop(s) with some willing and friendly volunteers. This allows you to build skills and confidence and iron out problems in your approach. Give them a topic which makes sense for them, not necessarily the one you will use in the real thing. Ask them to treat you firmly and offer critique and questions, you want to be pushed a bit, not just have a fun few hours with friends.

There are guides for workshops for individual methods, which tend to focus less on skills of facilitation and communication, and more on the technical details of a project or workshop or reflect on applying a method in a participatory mode. These are outlined in the individual methods chapters.

Now, get out there and start workshopping! We have found it a rewarding and interesting thing to do, you get to meet all sorts of different people, and you get a real feel for the value systems mapping can generate for people. You may be nervous, but you will enjoy it, promise!

References

Dialogue Matters. (2020). Better online meetings (Edition 2). Dialogue Matters Resource. https://dialoguematters.co.uk/wp-content/uploads/2020/05/Dialogue-Matters-Better-Online-Meetings-Edition-2.pdf

Fowler, B. (2020). Ten tips for facilitating online workshops. NPC blog. https://www.thinknpc.org/blog/ten-tips-for-facilitating-online-workshops/

Khuri, S., & Reed, M. (2020). *Tips and tools for making your online meetings and workshops more interactive*. Fast Track Impact Research Impact Guide. https://

www.fasttrackimpact.com/post/tips-and-tools-for-making-your-online-meetings-and-workshops-more-interactive

Penn, A. S., & Barbrook-Johnson, P. (2022). How to design a Participatory Systems Mapping process. CECAN toolkit. https://www.cecan.ac.uk/resources/toolkits/

CHAPTER 11

Comparing, Choosing, and Combining Systems Mapping Methods

Abstract This chapter explores (i) a detailed but usable comparison of the system mapping methods in this book; (ii) how we might choose which ones are appropriate given our and our project's needs, and the nature of the system we are working in; and (iii) how we might combine different sets of methods, both sequentially within a project, and in hybrid forms, to approach problems more holistically, and innovate methodologically.

Keywords Systems mapping • Method choice • Appropriateness

Making good comparisons and choosing which systems mapping methods to use is one of the most important things you will do in using systems mapping. This plays a big part in determining whether a participatory process is useful to stakeholders or not and whether a map can provide genuinely useful insights on the system of interest. Frequently, the choice also determines whether a good quality mapping process is even possible. Paradoxically, it is often done without much thought, if it is done at all. Often, we chose methods because they are what we know, have used before, or are asked for, based on rather random historical reasons. However, different methods have different strengths, weaknesses, and requirements, answer different sorts of questions, and work well in different situations. This makes it worthwhile broadening the range of methods in your toolkit and selecting them more thoughtfully.

P. Barbrook-Johnson, A. S. Penn, *Systems Mapping*,
https://doi.org/10.1007/978-3-031-01919-7_11

Helping researchers and practitioners to make better comparisons and choices is one of our main motivations for this book. Comparisons need to be made on solid grounds, with a good understanding of the variety of methods, of how maps and models are built but also in how they are analysed and used. We also need to make decisions with an appreciation of the flexibility of many methods, which means they can sometimes accommodate requirements that on face value may seem difficult to meet. This potentially infinite flexibility needs to be tempered by an understanding of where and when methods work best, and what their real strengths are. We want to be able to find the most appropriate method for our, and our stakeholders', needs, not just those that are adequate given several tweaks and adaptations.

The concept of 'appropriateness' is key here. It involves triangulating between the purpose, needs, and constraints of your project; the characteristics of the system and context you are working in; and the nature of the methods you are considering using. Where a method fits with both the project and the system, we can say it is an 'appropriate' method. This sounds simple in theory, and sometimes it is quite clear which methods are most appropriate; however, there are powerful forces which might interfere with our decision. We may have bias towards methods we know already, or methods which we are instinctively more comfortable with (e.g. if we are more of a quantitative or qualitative thinker), or methods we think others want us to use (e.g. funders, clients, or colleagues). We may feel stuck using one particular method which we have a track record with, which we get asked to use no matter the issue at hand. These biases and experiences will shape our choices of methods, but importantly, will also more fundamentally change the types of questions, topics, and domains we approach or interact in. It may be that we never put much thought into which methods to use because of the way our preferences and bias shape and narrow our thinking before we even get a chance to think about different methods or approaches.

Combining methods is not always necessary, but it is something we often do informally, as we adapt methods to our needs, taking inspiration from others. Methods can bleed into one another, that is, rather generic systems maps can be built that use some of the construction and analysis modes from multiple methods. For example, you might build a Causal Loop Diagram, but use some network analysis, or build an FCM where you really emphasise the feedbacks in the visualisation. If we had no time or budgetary constraints, combining methods is also something we would often recommend; to approach a question, topic, or issue from a variety of angles, to be more systematic or holistic in our thinking, to illuminate different aspects of a system, ask different questions or generate different perspectives. Exploratory combination of methods is a great source of innovation in methods and often a creative and rewarding endeavour. Thus, our default position is that combining methods in some form will normally be useful or valuable.

So, comparing and choosing systems mapping methods is important and sometimes difficult (i.e. if we feel constrained or don't have knowledge of multiple methods), but is also rarely done formally. And combining methods is an obvious way to improve our research and can be a creative and rewarding process. The rest of this chapter consider each of these steps in turn: we outline how we compare the methods in this book; consider how we might choose between methods, and outline some of the choices we might make; and consider what combinations, both sequentially and in hybrid form, might be interesting. Finally, we conclude with some tips on resources and getting started yourself.

COMPARING SYSTEMS MAPPING METHODS

In essence, the act of comparing methods is a simple task. All we need to do is use information on different methods, such as the information in this book, to compare the characteristics, pros, and cons of each method. In reality, each method chapter in this book is a significant simplification of what the method really is or can do. To make usable comparisons, we also have to summarise a lot of information, and variety, in a relatively small space—in our heads, or on a sheet of paper, or a slide. Table 11.1 attempts to do this 'usable' comparison for the methods in this book. We use the following categories to help us compare:

- **Type of map**: this captures the nature of the model that is created; whether it is quantitative or qualitative, the type of structure of the network used/allowed, and the way causality is represented (i.e. do connections imply direct causal influence or something more abstract).
- **Level of focus**: this draws attention to an often-forgotten characteristic of methods; they are not all focused on a whole system. Some focus on specific subsections of a system such as interventions and outcomes, or dynamical problems within them.

Table 11.1 Overview comparison of systems mapping methods in this book

Method	Level of focus	Type of map	Mode of construction	Emphasis on participation	Ease of use	Key contributions	Mode of analysis	Key constraints	When most appropriate
FCM	System	Semi-quantitative, cyclic, causal relations	Start with focal factors and build out, quantify connections (relative strengths)	High	Medium	'Quick and dirty' quasi-systems dynamics or relative causal importance of factors. Examination and comparison of stakeholder mental models	Compute impact of factors, rank, 'run' to equilibrium. Comparing network structures.	Analysis results highly sensitive to assumptions	When quick comparative semi-quantification wanted and flexibility needed in what map can include. When implications of assumptions, consolidation of inputs to relative size of impacts desired. When participation emphasised

| PSM | System | Qualitative, cyclic, causal relations | Start with focal factors and build out, collect info on factors and connections | High | Create bespoke analysis using submaps, network analysis, causal flow, and stakeholder info | Medium | Using submaps, network analysis, causal flow, and stakeholder information to make sense of large maps and providing 'actionable insights' | Difficult to share findings with those not involved with construction and analysis. No quantitative output | When participation emphasised, when flexibility of construction and analysis wanted, when large, inclusive maps desired |
| BBN | Intervention to outcomes within a system | Quantitative, acyclic, causal relations | Start with outcome and build back, or start with intervention and build down, define conditional probabilities | Neutral | Estimate effects of interventions, or contribution of factors to outcomes | Hard | Estimate impacts and contributions quantitatively. Quantification with low risk of producing meaningless analysis | Strong constraints on structure of map | When low-risk quantification wanted, when quantitative analysis of contributions to outcomes wanted |

(continued)

Table 11.1 (continued)

Method	Level of focus	Type of map	Mode of construction	Emphasis on participation	Mode of analysis	Ease of use	Key contributions	Key constraints	When most appropriate
CLD	System engine and feedbacks	Qualitative, cyclic, causal relations	Start with system engine, and feedback loops, build out, often use systems archetypes as templates/ inspiration.	Neutral	Develop qualitative understanding of potential dynamics and implications	Easy— medium	Intuitive focus on feedback loops and core system 'engine'. Gives qualitative sense of possible dynamics	Can produce very stylised version of a system focused on feedbacks. No quantitative output	When feedbacks or dynamical behaviour considered important, and quantification not required
SD	Dynamical problem within a system	Quantitative, cyclic, stock and flow, causal relations	Start with system engine, and feedback loops, build out and quantify	Neutral	Simulate dynamics through time	Hard	Simulate aggregate dynamics of a system through time, explore feedback loops quantitatively.	Most time consuming to use. Empirical validation often required. Sensitive to assumptions	When full rigorous simulation wanted, when feedbacks/ dynamical behaviour considered important
ToC	Intervention to outcomes within a system	Flexible, diagrammatic, cause and effect logic	Start with outcomes and intervention, and connect	Neutral	No formal analysis. Used to discuss assumptions, plan data collection.	Easy	Discipline intervention logic, plan intervention design and evaluation.	No analysis, may exclude important wider context	When focused on one intervention, when flexible 'framing tool' wanted

| RP | Situations within a system | Drawings, narrative | Free-form | High | Thematic | Easy | Inclusive and flexible understanding of system/situation. | Not a model in the same sense as other methods. Difficult to share findings with those not involved with construction and analysis. No quantitative output | When participation and rich expression are wanted |

- **Mode of construction**: this attempts to make clear the differences in how maps are started and built. This applies whether the method is being used in a participatory or a more data-driven mode. These differences are often missed, underplayed, or underappreciated by experts and beginners alike, but they can have a huge influence on the nature of the maps created.
- **Emphasis on participation**: this captures how much imperative is put on the need to use the maps in a participatory mode. Some methods foreground this, others are used in participatory and non-participatory modes.
- **Mode of analysis**: this captures the differences in how maps are analysed. These differences tend to be well understood by experts, but beginners don't always realise the range of ways we might analyse maps and the large differences in the types of insights that can offer.
- **Ease of use**: this attempts to describe how easy it is to use the methods with no formal training.
- **Key contributions and constraints**: these two categories only mention the most obvious or important contributions and constraints of the methods. There are others for all of the methods, but here we are trying to emphasise the most salient issues.
- **When most appropriate**: this is somewhat implicit from the other categories, but we use this category here to hammer-home the situations in which we think each method is at its best.

We hope the table is a useful tool for a relatively quick but detailed comparison of the methods we have focused on. You may want to use the table by adding new columns, extending the comparison into questions and topics specific to your problem, question, or project. Or you could add rows on other methods you are thinking of using. If anything is unclear or you want elaboration on any points, we suggest you start with the chapters on each method.

The history and roots of each method, and their deeper ontological and epistemological standpoints are missing from this table. We omit them here mainly because they are extremely difficult to meaningfully simplify down into table form. Nonetheless, the subtleties around a method's history, underlying assumptions, and philosophy are important. They are also often easy to misunderstand or underplay. It is common for important differences in methods' histories, which have affected how methods are thought about, framed, taught, and who has tended to use them, to go unremarked upon or underplayed. In each of the methods chapters in this book, we have attempted to give a sense of the history and philosophy of each method, so we strongly encourage readers to read these when looking at a method in detail, and to attempt to develop a sense of the history of a method.

Choosing Systems Mapping Methods

Choosing which method to use is more than just comparing their characteristics, pros, and cons. We need to also consider the aims, purpose, needs, and context of the project you are hoping to use them in, and the characteristics of the system you are studying. We need to triangulate between these three issues to find the method that is most appropriate. This will be the method that delivers the process, outputs, and insights the project needs in a timely and cost-effective fashion (given your and your users'/clients' resources), in forms which are usable for users, stakeholders, and clients, but which also captures the salient and important elements of the system we are working in. In practice, there are many interacting and competing elements to this decision, and we will often need to prioritise or weight these, and blend in our own preferences and expertise. Inspired by the design and choice triangles developed in Stern et al. (2012) and Befani (2020), we outline our take on this choice in Fig. 11.1.

It is possible to systematically approach this decision, categorising different requirements and characteristics, weighting and scoring them, to make a choice. Indeed, others have developed tools to help us do this, see Befani (2020) for example. However, in reality, few people make the decision in this way. We do not want to attempt to specify a detailed process for making the decision but rather encourage you to take time to reflect on it, really think about your needs and the characteristics of the system, as well as the methods yourself. Do think about the needs and constraints of the users and stakeholders of your research and how different methods will work for them.

You should also think about what data is available to you, or what mode you are thinking of using a method in (i.e. participatory, qualitative data, quantitative data, evidence review). It is technically true that any of the methods in this book can be used in any of these modes, and with or without any of these types of data. However, there are more and less common

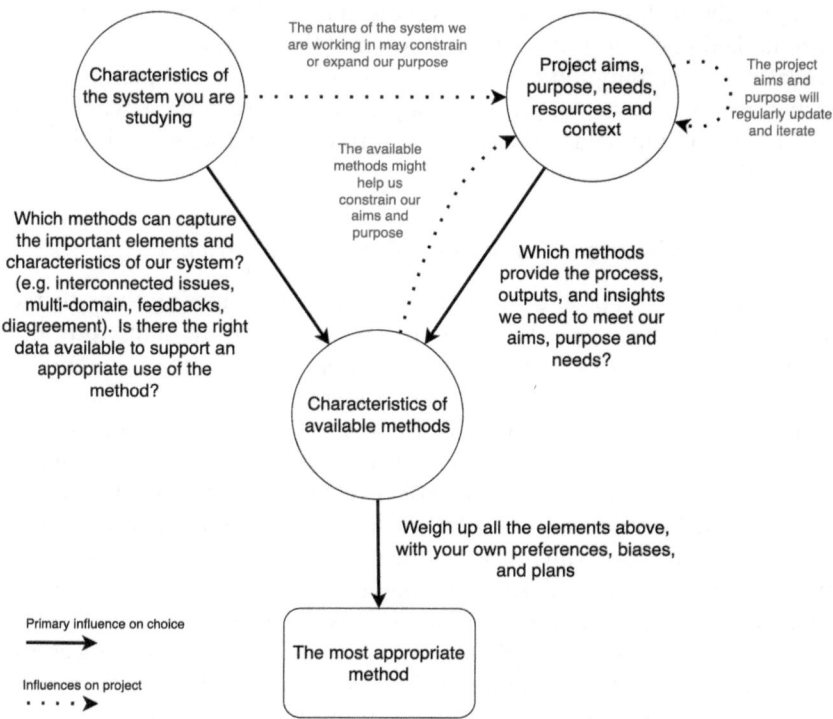

Fig. 11.1 Influences on the choice of most appropriate method. (Source: Authors' creation, inspired by Stern et al. (2012) and Befani (2020))

ways of using particular methods; for example, PSM is normally used without quantitative data, whereas System Dynamics models are often validated against quantitative data. In practice, people often feel that data availability constrains their modelling choices tightly. However, in reality, this is often only to fit into typical or accepted modes of using methods. We believe we should be more open about the way data availability constrains choice, but also be clearer about how that choice will constrain how model results can be interpreted and used.

Every situation will be different and have its idiosyncrasies, but to flesh this decision out a bit further, and build on the some of the information we summarised in the 'comparing' subsection above, in Table 11.2 we have sketched out the methods we think are most appropriate given

Table 11.2 Most appropriate methods given different project and system properties

Project aims, purpose, and needs	Most salient aspect of the system					
	Multi-domain system (i.e. social, economic, physical, biological factors)	Highly interconnected open system (i.e. many influential factors)	Feedbacks	Variety of stakeholder perspectives on the system	Dynamics (nonlinearity, path dependency, tipping points)	Network structure and leverage points
Understanding the system (from the researcher, user, client point of view)	PSM, CLD, ToC, FCM, BBN	PSM	CLD, SD, FCM	RP, FCM, PSM, ToC	SD, CLD	PSM, CLD, ToC, FCM, BBN
Forecasting	BBN	None	SD	None	SD	SD
Building consensus, groups, and capacity with stakeholders	RP, PSM, FCM	RP, PSM	CLD, FCM	RP, PSM	SD	PSM, ToC, FCM, BBN

different characteristics of a project and the system we are working in. It is important to note this table is a simplification and does not include all the aspects that will affect a decision, but it gives a feel for where methods work best.

Combining Systems Mapping Methods

Combining methods can be a powerful way to approach a question more holistically, generate novel insights, or innovate methodologically. It can also be a helpful way to bring in more people and expertise to a project, and to generate additional energy and enthusiasm around the analysis. Combining methods provides a natural mode of iteration, and way of expanding and increasing the depth of any project. Therefore, we think it worthwhile taking some time here to consider how we might combine the methods in this book, and we encourage you to think about how you could combine multiple methods to improve and extend your work. You may find it helpful to look at the large literatures on mixed methods (using qualitative and quantitative methods together) and multi methods (using two qualitative methods together) approaches (e.g. Anguera et al., 2018; Greene, 2008; Tashakkori & Teddlie, 2010); there is a long history there which is useful.

There are two basic ways to combine methods, either using them sequentially, with one method building on the outputs of the other, or directly together, in a hybrid method, where the aspects of two or more methods are directly brought together in one map. These two modes of combination are easily confused, using them sequentially means producing two mapping outputs, with one method using the other as an input. Hybridising them means producing one outputs that combines elements of two or more methods.

Figure 11.2 outlines some of the most obvious combinations we might make using two methods sequentially. In effect, these are workflows, of methods using the outputs of others. There are too many permutations of how methods might be more deeply hybridised, but Table 11.3 attempts to outline a few we believe are most likely to be valuable or interesting. This table is perhaps one of the best demonstrations in this book of the flexibility of these methods; we can relatively easily add aspects of other methods, either in the emphases we use in building maps, in annotations and additions we make to the full map diagram, and in ways we might analyse them.

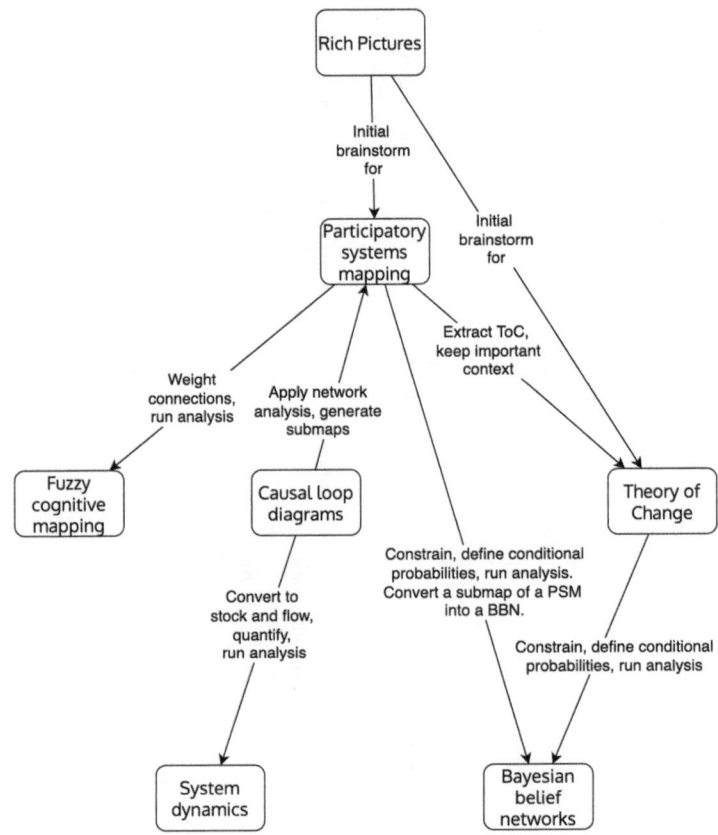

Fig. 11.2 Some of the potential sequential combinations of systems mapping methods. (Source: Authors' creation)

Brainstorming ways in which we might combine methods is quite fun (at least to us), but we need to make sure we are practical and rigours about it too. In any combination, there would be multiple of constraints, primarily because the building blocks of different mapping types, including the types of factors and connections and the connection structure allowed, may be fundamentally different. For example, we have suggested above the possibility of converting (parts of) a PSM into a BBN, which would require constraining the network (i.e. pruning connections so that there is maximum two inputs to a factor and removing any feedbacks). We

Table 11.3 Non-exhaustive list of potential hybridised elements of methods

	BBN	CLD	FCM	PSM	RP	SD	ToC
BBN				Do some network analysis	Annotate with RP-style drawing		Focus map on interventions and outcomes
CLD			Emphasise feedbacks in building	Apply network analysis and submaps	Annotate with RP-style drawing		Focus on interventions and outcomes
FCM		Emphasise feedbacks in building		Apply network analysis and submaps			Focus on interventions and outcomes
PSM		Emphasise feedbacks in building. Look for network archetypes in structure			Annotate with RP-style drawing		
RP		Embed map in diagram	Embed map in diagram	Embed map in diagram		Sketch plots of SD outputs	Embed map in diagram
SD				Apply network analysis and submaps			Focus on interventions and outcomes
ToC	Capture conditional dependencies of outcomes, based on inputs, activities etc.	Emphasise feedbacks in building		Capture more context. Apply PSM network analysis and submaps	Annotate with RP-style drawing		

Source: Authors' creation

Note: Read the table horizontally, that is, each row considers how the method on the left could be complemented by being combined with (an element of) the method in the column need to consider how we make the decisions on how to do this. Do we ask stakeholders to do it or do it ourselves? What is the basis for our judgements?

GETTING STARTED WITH CHOOSING AND COMBINING METHODS

Let's assume you don't need to do much more for comparing methods than reading this chapter, and some of the rest of this book (for the methods we cover at least). For choosing methods, there are some useful resources we would recommend:

- **Participatory modelling, 'selecting the right tool for the job' paper**: Voinov et al. (2018) gives an excellent overview of a range of participatory modelling methods and how we might go about choosing and combining them. It draws on a broader range of methods than just systems mapping, but still goes into a good level of detail and provides some case studies. The paper is long as it is worth taking the time to read it carefully, it is full of important nuggets of information.
- **Choosing appropriate evaluation methods**: Befani (2020) is focused on the evaluation of interventions but covers at least five of the methods (or similar methods) in this book. It provides extensive discussion on the factors to include in choosing methods and provides a well-tested spreadsheet tool for you to use too. This allows you to systematically score different aspects to identify and compare the appropriateness of methods.

On combining methods, there are not many resources out there which will directly help you think about how to combine systems mapping methods, it is inherently a creative process with many undefined spaces to explore, but you may find the following inspiring:

- **Mixing Operational Research methods**: Howick and Ackermann (2011) review examples of combinations of Operational Research methods, which includes some of the methods in this book (System Dynamics, different types of qualitative system models and influence maps, soft system methodology which includes Rich Pictures). They pull out some emerging themes and lessons from the combinations they consider, which are well worth reading.

- **Combining complexity-framed methods**: Barbrook-Johnson and Carrick (2021) review the combination of 'complexity-framed' methods, that is, methods which use the ideas and language of complexity science. They draw on a larger set of examples than Howick and Ackermann (2011), but necessarily conduct a lighter review of each example, looking for patterns in how and why combinations are made, before suggesting some potential combinations which are un(der)-explored.

For getting started yourself, most immediately, we would encourage you to take the 'comparing and choosing' decision more carefully than you might be tempted to. Don't rush it, or assume it is a no-brainer. Consider sketching out comparison tables applied to your context, to help you think through, and justify, your decision. Equally importantly, be sure to speak to the users and participants in your work and take their needs into consideration. If the model is not usable in their context, then it will just sit on a shelf gathering dust.

For combining methods, conceptually, the sky is the limit really, but it is hard to find the space and time to innovate methodologically or to approach projects from multiple angles. Thus, advocating, and making the time, for these activities may be a vital first step, before you even start to 'do it'. When you do, be creative and ambitious, but anchor your explorations in the same concerns as those you include when choosing methods and be sure that all the choices you make are underpinned in appropriate ways.

References

Anguera, M. T., Blanco-Villaseñor, A., Losada, J. L., Sánchez-Algarra, P., & Onwuegbuzie, A. J. (2018). Revisiting the difference between mixed methods and multimethods: Is it all in the name? *Quality & Quantity, 52*(6), 2757–2770. https://doi.org/10.1007/s11135-018-0700-2

Barbrook-Johnson, P., & Carrick, J. (2021). Combining complexity-framed research methods for social research. *International Journal of Social Research Methodology*. Online first. https://doi.org/10.1080/13645579.2021.1946118

Befani, B. (2020). Choosing appropriate evaluation methods: A tool for assessment and selection. *CECAN Report*. https://www.cecan.ac.uk/resources/toolkits/

Greene, J. C. (2008). Is mixed methods social inquiry a distinctive methodology? *Journal of Mixed Methods Research, 2*(1), 7–22. https://doi.org/10.1177/1558689807309969

Howick, S., & Ackermann, F. (2011). Mixing OR methods in practice: Past, present and future directions. *European Journal of Operational Research, 215*(3), 503–511. https://doi.org/10.1016/j.ejor.2011.03.013

Stern, E., Stame, N., Mayne, J., Forss, K., Davies, R., & Befani, B. (2012). Broadening the range of designs and methods for impact evaluations. *DFID Working Paper 38*. UK Department for International Development.

Tashakkori, A., & Teddlie, C. (2010). *SAGE handbook of mixed methods in social & behavioral research*. SAGE Publications, Inc. https://doi.org/10.4135/9781506335193

Voinov, A., Jenni, K., Gray, S., Kolagani, N., Glynn, P. D., Bommel, P., Prell, C., Zellner, M., Paolisso, M., Jordan, R., Sterling, E., Schmitt Olabisi, L., Giabbanelli, P. J., Sun, Z., Le Page, C., Elsawah, S., BenDor, T. K., Hubacek, K., Laursen, B. K., … Smajgl, A. (2018). Tools and methods in participatory modeling: Selecting the right tool for the job. *Environmental Modelling & Software, 109*, 232–255. https://doi.org/10.1016/j.envsoft.2018.08.028

CHAPTER 12

Conclusion

Abstract This chapter concludes the book by reflecting on what we have learnt writing it and presenting some of our more general takeaway lessons for your systems mapping practice.

Keywords Systems mapping • Modelling • Complexity • Policy • Systems

We hope you have found this book useful, despite its faults, and maybe even enjoyed bits of it! Ultimately, this is how we feel about systems mapping. We know that all models, and thus system maps, are not perfect; indeed, we should expect them to be wrong in some sense—they are simplifications of reality, with many judgements and assumptions used to cut the corners that create the simplification. They will have faults. We should not penalise systems mapping on this point. Rather, where we must be strict is in ensuring that system maps are useful. Either the process, the outputs, or both, of systems mapping must be useful for us and/or our stakeholders. It is this that you must keep forefront in your mind when designing and implementing systems mapping processes, and we hope this book will help you to do this.

Now indulge us in reflecting on what we have learnt by writing this book and some of our final few morsels of (subjective) advice.

P. Barbrook-Johnson, A. S. Penn, *Systems Mapping*,
https://doi.org/10.1007/978-3-031-01919-7_12

What Have We Learnt Writing This Book?

We have both learnt a huge amount preparing and writing this book; from one another, from our reading, and from interviews with experts on individual methods. Systems mapping, and its related methods, concepts, and schools of thought, is a messy space. There are multitudes of definitions and overlapping ideas, subtle differences in approach, and sweeping assumptions. We had underestimated this before; we did not know the waters were shark-infested! This makes it an exciting space to work in, with many different fields and approaches within touching distance, but also a difficult one to navigate. This can be frustrating and off-putting, but the quality, usefulness, and fun of using many of the methods here keeps pulling us back. We truly believe all the methods in this book are valuable in many circumstances, and are, on the whole, under-used in research and practice. We feel better placed to navigate this space having written this book, we hope you do too having read it.

Our Final Take-Home Messages

We have tried our best to be practical in how we have presented the methods in this book, but it is probably worthwhile rehearsing a few final take-away messages we encourage you to carry into your own work:

- **See system maps as living documents**: always try to build in a legacy for your work by having a plan for how your map can be maintained, updated, and re-used in the future. This need not be a fancy web-interface presenting your map but more often can be the capacity and skills you help create in a stakeholder team to continue the work themselves.
- **Be clear about whether you want value from the process, outputs, or both, of a mapping process**: value can come from the process of building a map (i.e. the discussions generated through building it, and the knowledge and mutual understanding this creates), from the analysis and use of the outputs of a process (i.e. the map itself, the narratives, insights, and knowledge it generates), or from both. You need to be clear on where you hope to get most value from systems mapping and manage expectations of stakeholders. We find we often expect much more value in process than the stakeholders we work with, who are often surprised just how much

they gain from this. The opportunity to reflect and generate under-standing with others is surprisingly powerful. Often more so than delivering 'answers'.

- **Climb the ladder of using methods**: imagine a short and stubby ladder with three steps, let us call this the 'ladder of using methods'. If you embark on climbing this ladder, first, you learn a method, and then want to apply it to everything and anything. You have your hammer, and now everything looks like a nail. On the second step, you become more aware of some similar methods to 'yours', perhaps ones you even used to see as competition, but now are more san-guine and maybe even know broadly when they should be used instead of 'your' method. On the third step, you have mastered some more methods, and now you can apply these too, but more impor-tantly, you now understand the detail and nuance of how each method operates and so can make informed decisions, with stake-holders, about what to use and when. We highly recommend you try to climb this ladder!
- **Decide whether you want to prioritise participation or concep-tual rigour**: in an ideal world we could do deeply participatory work which also produced models which were completely conceptually and scientifically sound. In practice, this is rarely possible. Stakeholders will want things included which a neutral 'expert' wouldn't, and they will often lead you off on tangents. This impedes your ability to make the model design decisions needed to make a model conceptu-ally and scientifically sound from a traditional modelling perspective. We need to acknowledge, accept, and plan for this, and again make sure our stakeholders understand this too.
- **Advocate for participatory steering of complex adaptive systems**: the philosophy that underpins much of our interest in systems map-ping is that we need ways of knowing and acting in the world, that are deeply participatory, holistic, and systemic in their scope, and humble in their attitude. We must accept that we cannot force or control the complex adaptive systems we live in and make up, but we can work with them to steer and nurture the change we want. We hope using systems mapping allows you to adopt this underlying worldview in your work.

Final Thoughts

Thank you for reading this book, and good luck with your systems mapping endeavours. We are always keen to hear about others' adventures, so do please get in touch!

INDEX

© The Author(s) 2022
P. Barbrook-Johnson, A. S. Penn, *Systems Mapping*,
https://doi.org/10.1007/978-3-031-01919-7